Practical Hydraulic Systems

Operation and Troubleshooting for Engineers and Technicians

Practical Hydraulic Systems

Operation and Troubleshooting for Engineers and Technicians

Ravi Doddannavar BEng (MechEng), Project manager and consultant, Pune, India

Andries Barnard Dip.MechEng, Senior Lecturer, Technikon Pretoria, South Africa

Series editor: Steve Mackay FIE (Aust), CPEng, BSc (ElecEng), BSc (Hons), MBA, Gov.Cert.Comp. Technical Director – IDC Technologies

ELSEVIER

AMSTERDAM • BOSTON • HEIDELBERG • LONDON
NEW YORK • OXFORD • PARIS • SAN DIEGO
SAN FRANCISCO • SINGAPORE • SYDNEY • TOKYO

Newnes is an imprint of Elsevier

Newnes

Newnes is an imprint of Elsevier
Linacre House, Jordan Hill, Oxford OX2 8DP, UK
30 Corporate Drive, Suite 400, Burlington, MA 01803, USA

First edition 2005
Reprinted 2008

British Library Cataloguing in Publication Data
A catalogue record for this book is available from the British Library

Library of Congress Cataloging-in-Publication Data
A catalog record for this book is available from the Library of Congress

ISBN: 978-0-7506-66276-5

For information on all Newnes publications
visit our website at www.newnespress.com

Transferred to Digital Printing in 2009.

Working together to grow
libraries in developing countries

www.elsevier.com | www.bookaid.org | www.sabre.org

ELSEVIER   BOOK AID
            International   Sabre Foundation

Contents

Preface

Whatever your hydraulic application, you can increase your knowledge of the fundamentals, improve your maintenance programs and become a more effective troubleshooter of problems in this area by reading this book. An attempt has been made to make the book practical and relevant. The areas of hydraulic systems construction, design, operations, maintenance and management issues are covered in this book.

Typical people who will hopefully find this book useful include:

- Plant engineers
- Operation, maintenance, inspection and repair managers, supervisors and engineers
- Mechanical engineers
- Design engineers
- Consulting engineers
- Plant operations and maintenance personnel
- Consulting engineers
- Process technicians
- Mechanical technicians.

We would hope that you will gain the following from this book:

- Ability to identify hydraulic systems components
- Knowledge of the essential hydraulic terms
- Ability to recognize the impact hydraulic fluids have on components
- Ability to describe the correct operation, control sequences and procedures for the safe operation of various simple hydraulic systems
- The knowledge to initiate an effective inspection and maintenance program.

You should have a modicum of mechanical knowledge and some exposure to industrial hydraulic systems to derive maximum benefit from this book.

1

Introduction to hydraulics

1.1   Objectives

Upon completing this chapter, one should be able to:

- Understand the background and history of the subject of hydraulics
- Explain the primary hydraulic fluid functions and also learn about the basic hydraulic fluid properties
- Understand how important fluid properties like velocity, acceleration, force and energy are related to each other, and also learn about their importance in relation to hydraulic fluids
- Understand the concepts of viscosity and the viscosity index
- Explain the lubrication properties of a hydraulic fluid.

1.2   Introduction and background

In the modern world of today, hydraulics plays a very important role in the day-to-day lives of people. Its importance can be gaged from the fact that it is considered to be one part of the muscle that moves the industry, the other being Pneumatics. The purpose of this book is to familiarize one with the underlying principles of hydraulics as well as make an effort at understanding the practical concepts governing the design and construction of various hydraulic systems and their applications. Additionally the functional aspects concerning the main hydraulic system components as well as the accessory components have been dealt with, in detail. The final part of the book is devoted to the general maintenance practices and troubleshooting techniques used in hydraulic systems with specific emphasis on ways and means adopted to prevent component/system failures.

The Greek word 'Hydra' refers to water while 'Aulos' means pipes. The word hydraulics originated from Greek by combining these words, which in simple English means, *water in pipes*. Man has been aware of the importance of hydraulics since prehistoric times. In fact even as early as the time period between 100 and 200 BC, man had realized the energy potential in the flowing water of a river. The principles of hydraulics were put to use even in those early times, in converting the energy of flowing water into useful mechanical energy by means of a water wheel.

Ancient historical accounts show that water was used for centuries to generate power by means of water wheels. However, this early use of fluid power required the movement of huge quantities of fluid because of the relatively low pressures provided by nature.

With the passage of time, the science of hydraulics kept on developing as more and more efficient ways of converting hydraulic energy into useful work were discovered. The subject of hydraulics which dealt with the physical behavior of water at rest or in motion remained a part of civil engineering for a long time. However, after the invention of James Watt's 'steam engine', there arose the need for efficient transmission of power, from the point of generation to the point of use. Gradually many types of mechanical devices such as the line shaft, gearing systems, pulleys and chains were discovered. It was then that the concept of transmitting power through fluids under pressure was thought of. This indeed was a new field of hydraulics, encompassing varying subjects such as power transmission and control of mechanical motion, while also dealing with the characteristics of fluids under pressure.

To distinguish this branch of hydraulics from water hydraulics, a new name called 'Industrial hydraulics' or more commonly, 'oil hydraulics' was coined. The significance behind choosing this name lies in the fact that this field of hydraulics employs oil as a medium of power transmission. Water which is considered to be practically incompressible is still used in present-day hydrotechnology. The term *water hydraulics* has since been coined for this area of engineering. But by virtue of their superior qualities such as resistance to corrosion as well as their sliding and lubricating capacity, oils which are generally mineral-based are the preferred medium for transmission of hydraulic power.

The study of 'Oil Hydraulics' actually started in the late seventeenth century when Pascal discovered a law that formed the fundamental basis for the whole science of hydraulics. The concept of undiminished transmission of pressure in a confined body of fluid was made known through this principle. Later Joseph Bramah, developed an apparatus based on Pascal's law, known as *Bramah's press* while Bernoulli developed his law of conservation of energy for a fluid flowing in a pipeline. This along with Pascal's law operates at the very heart of all fluid power applications and is used for the purpose of analysis, although they could actually be applied to industry only after the industrial revolution of 1850 in Britain.

Later developments resulted in the use of a network of high-pressure water pipes, between generating stations having steam-driven pumps and mills requiring power. In doing this, some auxiliary devices such as control valves, accumulators and seals were also invented. However, this project had to be shelved because of primarily two reasons, one the non-availability of different hydraulic components and two, the rapid development of electricity, which was found to be more convenient and suitable for use.

A few developments towards the late nineteenth century led to the emergence of electricity as a dominant technology resulting in a shift in focus, away from fluid power. Electrical power was soon found to be superior to hydraulics for transmitting power over long distances.

The early twentieth century witnessed the emergence of the modern era of fluid power with the hydraulic system replacing electrical systems that were meant for elevating and controlling guns on the battleship USS Virginia. This application used oil instead of water. This indeed was a significant milestone in the rebirth of fluid power hydraulics. After World War II, the field of hydraulic power development has witnessed enormous development. In modern times, a great majority of machines working on the principle of 'oil hydraulics' have been employed for power transmission. These have successfully been able to replace mechanical and electrical drives. Hydraulics has thus come to mean, 'the science of the physical behavior of fluids'.

1.3   Classification

Any device operated by a hydraulic fluid may be called a hydraulic device, but a distinction has to be made between the devices which utilize the impact or momentum of a moving fluid and those operated by a thrust on a confined fluid i.e. by pressure. This leads us to the subsequent categorization of the field of hydraulics into:

- Hydrodynamics and
- Hydrostatics.

Hydrodynamics deals with the characteristics of a liquid in motion, especially when the liquid impacts on an object and releases a part of its energy to do some useful work.

Hydrostatics deals with the potential energy available when a liquid is confined and pressurized. This potential energy also known as hydrostatic energy is applied in most of the hydraulic systems. This field of hydraulics is governed by Pascal's law.

It can thus be concluded that pressure energy is converted into mechanical motion in a hydrostatic device whereas kinetic energy is converted into mechanical energy in a hydrodynamic device.

1.4   Properties of hydraulic fluids

The single most important material in a hydraulic system is the working fluid itself. Hydraulic fluid characteristics have a major influence on the equipment performance and life and it is therefore important to use a clean high-quality fluid so that an efficient hydraulic system operation is achieved. Essentially, a hydraulic fluid has four primary functions:

1. *Transmission of power*:   The incompressibility property of the fluid due to which energy transfer takes place from the input side to the output side (Figure 1.1).

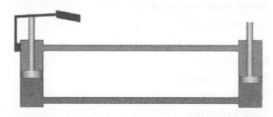

Figure 1.1
Energy transfer property of a hydraulic fluid

2. *Lubrication of moving parts*:   Lubrication function of the fluid minimizes friction and wear (Figure 1.2).

Figure 1.2
Lubrication property of a hydraulic fluid

3. *Sealing of clearances between mating parts*:   The fluid between the piston and the wall acts as sealant (Figure 1.3).

Figure 1.3
Sealing property of a hydraulic fluid

4. *Dissipation of heat*:   Heat dissipation due to the heat transfer property of the hydraulic fluid (Figure 1.4).

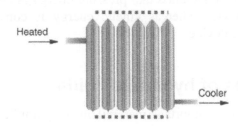

Figure 1.4
Heat transfer property of a hydraulic fluid

For the hydraulic fluid to properly accomplish these primary functions, the following properties are quite essential:

- Good lubricity
- Ideal viscosity
- Chemical and environmental stability
- Large bulk modulus
- Fire resistance
- Good heat transfer capability
- Low density
- Foam resistance
- Non-toxicity
- Low volatility.

Last but not the least, the fluid selected must be cost-effective and readily available. It is quite obvious that a clear understanding of the fundamentals of fluids is required to fully comprehend the concepts of hydraulics. We shall therefore briefly review certain important terms and definitions that are often used in hydraulics.

1.4.1   Fluids

A liquid is a fluid, which for a given mass will have a definite volume independent of the shape of its container. This implies that the liquid will fill only that part of the container whose volume equals the volume of the liquid although it assumes the shape of the container. For example, if we pour water into a vessel and the volume of water is not sufficient to fill the vessel, then a free surface (Figure 1.5) will be formed as shown in Figure 1.1.

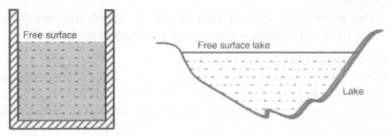

Figure 1.5
Free surface of a liquid

Unlike gases, liquids are hardly compressible and that is the reason why their volume does not vary with change in pressure. Though this is not completely true as changes in volume do occur on account of variations in pressure, these changes are so small that they are at best ignored for most engineering applications.

Gases on the other hand are fluids that are easily compressible. Therefore unlike liquids which have a definite volume for a given mass, the volume of a given mass of a gas will increase in order to fill the vessel that contains the gas. Furthermore, gases are greatly influenced by the pressure to which they are subjected. An increase in pressure causes the volume of the gas to decrease and vice versa. Air is the only gas commonly used in hydraulic systems because it is inexpensive and readily available.

1.4.2 Mass

The mass of a body or an object is a measure of the quantity of matter contained in it. The mass of a body is constant and independent of the surroundings and position. A physical balance is used to measure the mass of a body. Mass is normally measured in kilograms (kg) or in pounds (lbs). The mass of 1 liter of water at 4 °C is taken as 1 kg. The other commonly used unit of mass is the metric ton, where 1 metric ton = 1000 kg.

1.4.3 Volume

The space occupied by a body is called its volume. Volume is usually expressed in terms of cubic meters (m^3) or cubic feet (ft^3) or liters. One liter is equal to 1000 cm^3 and is equal to the volume of 1 kg of water at 4 °C.

The units of volume are related as follows:

$1 \ m^3 = 1000 \ \text{liter}$
$1 \ dm^3 = 1000 \ cm^3 = 1 \ \text{liter}$
$1 \ cm^3 = 1 \ ml = 1000 \ mm^3$

1.4.4 Density

The density of a substance is defined as its mass per unit volume. It is denoted by the symbol 'ρ' (rho). If equal masses of cotton and lead are taken (say 1 kg each), we will find that the volume of cotton is much larger than the volume of lead. This is because lead is heavier (denser) than cotton. The particles of lead are closely packed while those of cotton are more diffused.

Density for a given substance can be calculated from the following equation:

$$\text{Density} \ (\rho) = \frac{\text{Mass of the substance} \ (m)}{\text{Volume of the substance} \ (V)}$$

The mass of 1 cm^3 of iron is 7.8 g; hence the density of iron is 7.8 g/cm^3 or 7.8 × 10^3 kg/m^3. Density changes with change in temperature.

For example:

When water is cooled to 4 °C, it contracts i.e. its volume decreases, thereby resulting in an increase in density. But if water is further cooled below 4 °C, it begins to expand i.e. its volume increases and hence its density decreases. Thus, the density of water is a maximum at 4 °C and is 1 gm/cm^3 or 1000 kg/m^3.

1.4.5   Relative density or specific gravity

The relative density of a substance is the ratio of its density to the density of some standard substance. It is denoted by the letter 's'. The standard substance is usually water (at 4 °C) for liquids and solids, while for gases it is usually air.

$$\text{Relative density for liquids and solids } (s) = \frac{\text{Density of substance}}{\text{Density of water at 4°C}}$$

$$\text{Relative density for gases } (s) = \frac{\text{Density of substance}}{\text{Density of air}}$$

Density of substance (liquid or solid) = Density of water at 4 °C × Relative density of the substance i.e. ρ (solids and liquids) = $1000 \times s$ and ρ (gases) = $1.29 \times s$

Since 'relative density' is a pure ratio, it has no units.

1.4.6   Velocity

The distance covered by a body in a unit time interval and in a specified direction is called velocity.

If the body travels equal distances in equal intervals of time along a particular direction, the body is said to be moving with a uniform velocity.

If the body travels unequal distances in a particular direction at equal intervals of time or if the body moves equal distances in equal intervals of time but with a change in its direction, the velocity of the body is said to be variable.

$$\text{The average velocity } (v) = \frac{\text{Total distance traveled in a specific direction } (S)}{\text{Total time of travel } (t)}$$

The unit of velocity is meters/second (m/s) or kilometers/hour (km/h).

1.4.7   Acceleration

Generally, bodies do not move with constant velocities. The velocity may change in either magnitude or direction or both. For example, consider a car changing its speed while moving in a busy street. This leads us to the concept of acceleration which may be defined as the rate of change of velocity of a moving body.

The acceleration is said to be uniform when equal changes in velocity take place in equal intervals of time, however small these intervals may be. If the velocity is increasing, the acceleration is considered as positive. If the velocity is decreasing, the acceleration is negative and is usually called deceleration or retardation.

$$\text{Acceleration } (a) = \frac{\text{Final velocity } (v_f) - \text{Initial velocity } (v_o)}{\text{Time interval over which the change occurred } (t)}$$

The units of acceleration are ft/s^2 or m/s^2.

1.4.8 Acceleration due to gravity

The acceleration produced by a body falling freely under gravity due to the earth's attraction is called 'acceleration due to gravity'. It is denoted by the letter 'g'.

If a body falls downwards, the acceleration due to gravity is said to be positive, while if the body moves vertically upwards, the acceleration is said to be negative. The average value of acceleration due to gravity is 9.8 m/s^2 (approximately 32 ft/s^2). Thus for a freely falling body under gravity, its velocity increases at the rate of 9.8 m/s i.e. after 1 s the velocity will be 9.8 m/s, after 2 s the velocity will be $9.8 \times 2 = 19.6$ m/s and so on.

Actually, the value of 'g' varies from place to place. On the earth's surface, 'g' is said to be maximum at the poles and minimum at the equator.

1.4.9 Force

Consider the following:

- The pushing of a door to open it
- The pulling of a luggage trolley
- The stretching of a spring by a load suspended on it.

In the above examples, we have a force exerting a push, pull or stretch. The magnitude of the force is different in each case and is dependent on the size and content of the object.

The force in the above cases is called the 'force of contact' because the force is applied by direct contact with the body. The force is either changing the position/displacement of the object or its dimensions. The magnitude of the force due to gravity on an object depends upon the mass of the object.

At any given place, the force of gravity is directly proportional to the mass of the body. The force due to gravity on a mass of 1 kg is called a 1 kg force (1 kgf) or if expressed in terms of Newton, 9.8 Newton.

It can be derived experimentally that if, a force (F) acts on an object of mass (m), the object accelerates in the direction of the force. The acceleration (a) is proportional to the force and inversely proportional to the mass of the object.

$$F = m\,a$$

This relationship is also referred to as Newton's second law of motion.

As discussed above, in the SI system, the unit of force is 'Newton' which is abbreviated as N. One Newton is defined as that force which while acting on a body of mass 1 kg, produces an acceleration of 1 m/s^2.

1.4.10 Weight

Weight refers to the force of gravity acting on a given mass.

On the earth, weight is the gravitational force with which the earth attracts the object. If 'm' is the mass of the object, then the weight is given by the relationship,

Weight (W) of the object = Mass of the object (m) × acceleration due to gravity (g)

So,

$$W = m \times g$$

The unit of weight (in SI units) is Newton (N). Since 'g' on earth is 9.81 m/s², a 1 kgf object weighs 9.8 N on earth.

$$1 \text{ kgf} = 9.81 \text{ N}$$

1.4.11 Specific weight

The specific weight or weight density of a fluid is defined as the ratio of the weight of the fluid to its volume. It is denoted by the letter 'w'. Thus the weight per unit volume of a fluid is called the weight density.

$$
\begin{aligned}
\text{Weight density} &= \frac{\text{Weight of the fluid}}{\text{Volume of the fluid}} \\
&= \frac{\text{Mass of fluid } (m) \times (g)}{V}
\end{aligned}
$$

Since m/V is density (ρ), the equation for weight can be written as

$$w = \rho \times g$$

So, weight density (w) = mass density (ρ) × acceleration due to gravity (g)
Specific weight of water is given by = $1000 \times 9.81 = 9810$ N/m³ (in SI units).

1.4.12 Work

Work is defined as force through distance. In other words, when a body moves under the influence of a force, work is said to be done. On the contrary, if there is no motion produced on the body, the work done is zero. Thus work is said to be done only when the force is applied to a body to make it move (i.e. there is displacement of the body). If you try to push a heavy boulder but you are unable to get it to move, then the work done will be zero. Referring to Figure 1.6, work is said to be accomplished if we move 100 kg a distance of 2 m. The amount of work here is measured in kg m.

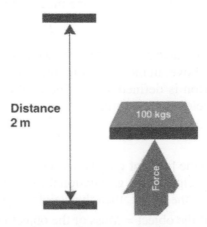

Figure 1.6
Principle of work

The work done will be large, if the force required to displace the body is large or if the displacement of the body due to the applied force is large. The mathematical formula to calculate the work done is

$$\text{Work } (W) = \text{Force } (F) \times \text{Distance moved or displacement } (s)$$
$$W = Fs$$

The SI unit of work is Newton-meters which is also referred to as joules (J). One joule is the work done by a force of 1 N when it displaces a body by 1 m in the direction of the force.

1.4.13 Energy

A body is said to possess energy when it is capable of doing work. Therefore, energy may be broadly defined as the ability to do work. In other words, energy is the capacity of a body for producing an effect. In hydraulics, the method by which energy is transferred is known as fluid power. The energy transfer takes place from a prime mover or input power source to an output device or actuator.

Energy is further classified as:

- *Stored energy*:   Examples being chemical energy in fuel and energy stored in water.
- *Energy in transition*:   Examples being heat and work.

The following are the various forms of energy:

Potential energy (PE)

It is the energy stored in the system due to its position in the gravitational field. If a heavy object such as a large stone is lifted from the ground to the roof, the energy required to lift the stone is stored in it as potential energy. This stored potential energy remains unchanged as long as the stone remains in its position.

Potential energy is given by

$$\text{PE} = z \times g$$

Where z is the height of the object above the datum.

Kinetic energy (KE)

Kinetic energy is the energy possessed by a body by virtue of its motion. If a body weighing 1 kg is moving at a velocity of v m/s with respect to the observer, then the kinetic energy stored in the body is given by:

$$\text{KE} = \frac{v^2}{2}$$

This energy will remain stored in the body as long as it continues in motion at a constant velocity. When the velocity is zero, the kinetic energy is also zero.

Internal energy

Molecules possess mass and have both translational and rotational motion in liquid and gaseous states. Owing to both, their mass as well as their motion, these molecules have a large amount of kinetic energy stored in them. Any change in the temperature results in a change in the molecular kinetic energy, since molecular velocity is a function of temperature.

In addition, the molecules in the solid state are attracted towards each other by forces, which are quite large. These forces tend to vanish once the molecules attain a perfect gas state. In processes such as melting of a solid or vaporization of a liquid, it is necessary to overcome these forces. The energy required to bring about this change is stored in the molecules as potential energy.

The sum of these energies is called internal energy, and is stored within the body. We refer to this energy as internal energy or thermal energy denoted by the symbol 'u'.

Energy is usually expressed in terms of British thermal unit (Btu) or joule (J).

1.4.14   Power

The rate of doing work is called power. It is measured as the amount of work done in 1 s. If the total work done in time 't' is 'W' then

$$\text{Power } (P) = \frac{\text{Work done } (W)}{\text{Time } (t)}$$

This can be written as

$$\text{Power} = \text{Force} \times \text{Average velocity}$$
$$P = F \times v$$

Since work done = force × distance and velocity = distance/time.

From Figure 1.7, if we lift 100 kg, 2 m in 2 s, we have accomplished 100 units of power or in other words, 100 times 2 divided by 2 s. This is usually converted into kilowatt or horsepower in order to obtain a relative meaning for measuring power.

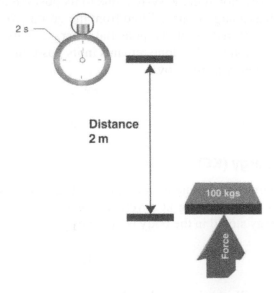

Figure 1.7
Principle of power

The SI unit of power is J/s or W. If the amount of work done is 1 J in 1 s, than the power will be 1 W.

$$\therefore 1\,\text{W} = 1\,\text{J/s}$$

Larger units of power are kilowatts (kW) and Megawatts (MW).

$$1\,\text{kW} = 1000\,\text{W}$$
$$1\,\text{MW} = 10^6\,\text{W}$$

The practical unit of power that is often used in mechanical engineering is horsepower (hp).

Horsepower

A horsepower is the power of one horse, or a measure of the rate at which a single horse can work. When we specify an engine as 30 hp, it implies that the engine can do the work of 30 horses.

One horse is said to be capable of walking 50 m in 1 min, lifting a 90 kgf weight.

$$\text{Work done by the horse} = 90 \times 50 = 4500\,\text{kgf m}$$
$$\text{Power} = \text{Work done/time}$$
$$= 4500\,\text{kgf m/min}$$
$$1\,\text{hp} = 4500/60 = 75\,\text{kgf m/s.}$$
$$1\,\text{hp} = 746\,\text{W}$$

We have mentioned earlier that energy is expressed in a larger unit called kilowatt-hour (kWh).

$$1\,\text{kWh} = 1\,\text{kW} \times 1\,\text{h}$$
$$= 1000\,\text{J/s} \times 60 \times 60\,\text{s}$$
$$= 3.6 \times 10^6\,\text{J}$$
$$1\,\text{kWh} = 3.6 \times 10^6\,\text{J}$$
$$1\,\text{Wh} = 3.6 \times 10^3\,\text{J}$$

1.4.15   Bulk modulus

The highly favorable power to weight ratio and their stiffness in comparison with other systems makes hydraulic systems an obvious choice for high-power applications. The stiffness of a hydraulic system is directly related to the incompressibility of the oil. Bulk modulus is a measure of this compressibility. Higher the bulk modulus, the less compressible or stiffer is the fluid.

The bulk modulus is given by the following equation:

$$\beta = -V\left(\frac{\Delta P}{\Delta V}\right)$$

Where

V is the original volume

ΔP is the change in pressure and

ΔV is the change in volume.

1.4.16   Viscosity and viscosity index

Viscosity is considered to be probably the single most important property of a hydraulic fluid. It is a measure of the sluggishness at which the fluid flows or in other words a measure of a liquid's resistance to flow. A thicker fluid has higher viscosity and thereby increased resistance to flow. Viscosity is measured by the rate at which the fluid resists deformation. The viscosity property of the fluid is affected by temperature. An increase in the temperature of a hydraulic fluid results in a decrease in its viscosity or resistance to flow.

Too high a viscosity results in:

- Higher resistance to flow causing sluggish operation
- Increase in power consumption due to frictional losses
- Increased pressure drop through valves and lines
- High temperature conditions caused due to friction.

Too low a viscosity results in:

- Increased losses in the form of seal leakage
- Excessive wear and tear of the moving parts.

Viscosity can be further classified as:

- Absolute viscosity and
- Kinematic viscosity.

Absolute viscosity   Also known as the coefficient of dynamic viscosity, absolute viscosity is the tangential force on a unit area of either one or two parallel planes at a unit distance apart when the space is filled with liquid and one of the planes moves relative to the other at unit velocity. It is measured in poise. The most commonly used unit is Centipoise, which is 1/100th of a poise.

Kinematic viscosity   Most of the calculations in hydraulics involve the use of kinematic viscosity rather than absolute viscosity. Kinematic viscosity is a measure of the time required for a fixed amount of oil to flow through a capillary tube under the force of gravity. It can also be defined as the quotient of absolute viscosity in centipoise divided by the mass density of the fluid. Kinematic viscosity can be mathematically represented as $v = \mu/\rho$. It is usually measured in centistokes. The viscosity of a fluid is measured by a Say bolt viscometer, whose schematic representation is shown in Figure 1.8.

This device consists of an inner chamber containing the oil sample to be tested. A separate outer compartment, which surrounds the inner chamber, contains a quantity of oil whose temperature is controlled by a thermostat and a heater. A standard orifice is located at the bottom of the center oil chamber. When the oil attains the desired temperature, the time it takes to fill up a 60 cm^3 container through the metering orifice is recorded. The time (t) measured in seconds is the viscosity in Saybolt universal seconds (SUS). The SUS viscosity for a thick fluid will be higher than that for a thin fluid, since it flows slowly.

To convert SUS to centistokes, the following empirical equations are used,

$$v \text{ (centistokes)} = \frac{0.226\,t - 195}{t}, \quad \text{for} \quad t \leq 100\,\text{SUS} \quad \text{and}$$

$$v \text{ (centistokes)} = \frac{0.220\,t - 135}{t}, \quad \text{for} \quad t > 100\,\text{SUS}$$

Where v represents the viscosity in centistokes and the time measured in SUS or simply seconds.

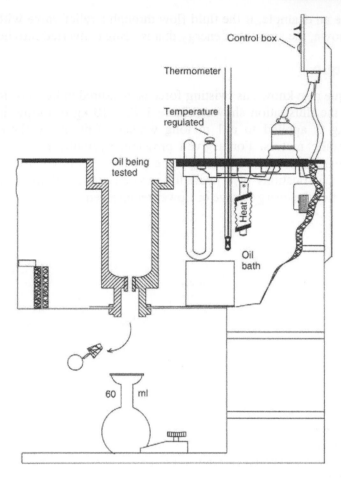

Figure 1.8
Say bolt viscometer

Viscosity index

The viscosity index is an empirical number indicating the rate of change of viscosity of an oil within a given temperature range. A low viscosity index indicates a relatively larger change in viscosity with temperature whereas a high viscosity index indicates a relatively smaller change in viscosity with temperature.

The viscosity index is calculated as follows:

$$V I = \frac{(L - U)}{(L - H)} \times 100$$

Where
   U is the viscosity in SUS of the oil whose viscosity index is to be calculated at 37.8 °C or 100 °F
   L is viscosity in SUS of the oil of '0' viscosity index at 37.8 °C (100 °F) and
   H is the viscosity in SUS of the oil of '100' viscosity index at 37.8 °C (100 °F).

1.4.17   Heat

This is another important property associated with hydraulic fluids. According to the law of conservation of energy, although heat undergoes a change in form, it can neither be created nor destroyed. The unused energy in a hydraulic system takes the form of heat. To quote an example, if the fluid flow through a relief valve with a standard pressure setting is known, the amount of energy that is being converted into heat can be easily calculated.

1.4.18   Torque

Torque also known as twisting force is measured in kg-m or foot-pounds.

In the illustration shown (Figure 1.9), a 10 kg-m torque is produced when a force of 10 kg is applied to a 1 m long wrench. This is the theory that finds application in hydraulic motors. For a given pressure, hydraulic motors are rated at specific torque values. The torque or twisting force produced in a hydraulic motor is the generated work. The specifications of a hydraulic motor in terms of its rpm at a given torque capacity specifies the energy usage or power requirement.

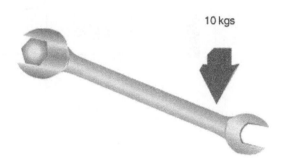

Torque = 10 kg-m

10 kgs

Figure 1.9
Principle of torque

1.4.19   Lubrication

Hydraulic fluids should have good lubrication properties to prevent wear and tear between the closely fitting moving parts. Direct metal-to-metal contact of the hydraulic components is normally avoided by employing fluids having adequate viscosity which tend to form a lubricating film between the moving parts (Figure 1.10). This has been illustrated in Figure 1.3.

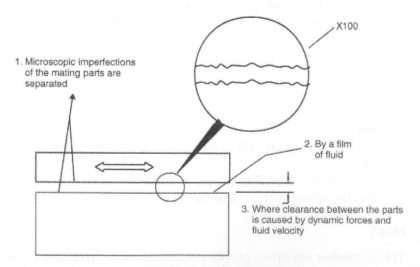

1. Microscopic imperfections of the mating parts are separated

X100

2. By a film of fluid

3. Where clearance between the parts is caused by dynamic forces and fluid velocity

Figure 1.10
Lubricating film prevents metal-to-metal contact

The hydraulic components that suffer the most from conditions arising out of inadequate lubrication include pump vanes, valve spools, rings and rod bearings.

Wear and tear is the removal of surface material due to the frictional force between two mating surfaces. It has been determined that the frictional force is proportional to the normal force which forces the two surfaces together and the proportionality constant is known as the coefficient of friction (CF).

2

Pressure and flow

2.1 Objectives

On reading this chapter, the student will be able to:

- Explain and understand the various terms and definitions used in hydraulics
- Understand the significance of Pascal's law and its applications
- Understand the importance of flow and pressure in hydraulics.

2.2 Pressure

Pressure along with flow is one of the key parameters involved in the study of hydraulics. Pressure in a hydraulic system comes from resistance to flow. This can be best understood from Figure 2.1.

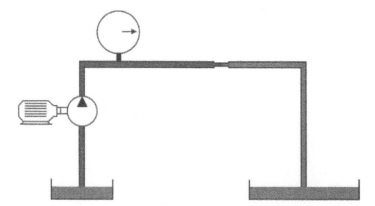

Figure 2.1
Pressure buildup in a hydraulic system

Consider the flow from a hydraulic pump as shown. Here the pump produces only flow and not pressure. However any restriction in the flow from the pump results in the formation of pressure. This restriction or resistance to flow normally results from the load induced in the actuator. The various conductors and components of the hydraulic system such as pipes and elbows also act as points of resistance and contribute to the generation of pressure in the system.

Pressure (P) is defined as the force (F) acting normally per unit area (A) of the surface and is given by the equation:

$$P = \frac{F}{A}$$

Pressure in the SI unit is measured in terms of N/m^2 also known as a Pascal. Pressure can also be expressed in terms of bar, where

$$1 \text{ bar} = 10^5 \text{ N/m}^2$$

Pressure in the US unit is measured in terms of lb/in.^{2} or psi, where

$$1 \text{ psi} = 0.0703 \text{ kg/cm}^2$$

2.2.1   Pressure in fluids

Fluids are composed of molecules, which are in continuous random motion. These molecules move throughout the volume of the fluid colliding with each other and with the walls of the container as a result of which the molecules undergo a change in momentum.

Now, let us consider a surface within the fluid which is impacted by a large number of molecules. This results in a transfer in momentum from the molecules to the surface. The change in momentum transferred per second by these molecules on the surface gives the average force on the surface, while the normal force exerted by the fluid per unit area of the surface is known as fluid pressure.

2.2.2   Pressure at a point in a liquid

The pressure at any point in a fluid at rest, is given by the Hydrostatic law, which states that the rate of increase of pressure in a vertically downward direction must be equal to the specific weight of the fluid at that point.

The vertical height of the free surface above any point in a liquid at rest is known as the pressure head. This implies that the pressure (called head pressure) at any point in a liquid is given by the equation:

$$P = \rho g h$$

Where
ρ is the density of the liquid
h is the free height of the liquid above the point and
g is the acceleration due to gravity.

Thus, the pressure at any point in a liquid is dependent on three factors:

1. Depth of the point from the free surface
2. Density of the liquid
3. Acceleration due to gravity.

2.2.3   Atmospheric, absolute, gage pressure and vacuum

Atmospheric pressure

The earth is surrounded by an envelope of air called the atmosphere, which extends upwards from the surface of the earth. Air has mass and due to the effect of gravity exerts a force called weight. The force per unit area is called pressure. This pressure exerted on the earth's surface is known as atmospheric pressure.

Gage pressure

Most pressure-measuring instruments measure the difference between the pressure of a fluid and the atmospheric pressure. This is referred to as gage pressure.

Absolute pressure

Absolute pressure is the sum of the gage pressure and the atmospheric pressure.

Vacuum

If the pressure is lower than the atmospheric pressure, its gage pressure is negative and the term vacuum is used when the absolute pressure is zero (i.e. there is no air present whatsoever).

$$P = 0$$

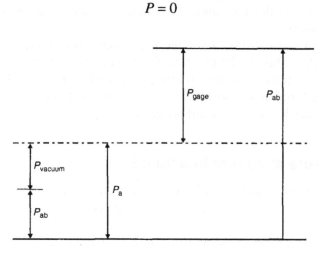

Figure 2.2
Relationship between absolute, gage and vacuum pressure

In Figure 2.2, P_a is the atmospheric pressure, P_{gage} is the gage pressure, P_{ab} is the absolute pressure and P_{vacuum} is the vacuum pressure.

2.2.4   Effect of pressure on boiling point

The boiling point of a liquid increases with an increase in pressure while conversely decreasing with a decrease in pressure. Thus if the atmospheric pressure is more than 14.7 psi or 101.3 kPa, water boils at a temperature higher than 100 °C (212 °F). Similarly water boils at a lower temperature if the pressure is lower than 14.7 psi or 101.3 kPa.

At boiling point, the pressure of the vapors at the liquid surface is equal to the external atmospheric pressure. Thus if the external atmospheric pressure increases, the liquid has to boil at a higher temperature to create a vapor pressure equal to the external pressure.

At higher altitudes, the atmospheric pressure is low, hence water boils at a temperature lower than 100 °C (212 °F). This makes cooking difficult. An important point to be noted is that adding impurities to a liquid can increase its boiling point.

2.2.5   Pressure measurement

The behavior of a fluid can be deduced by measuring the two critical system parameters of flow and pressure. For flow measurement, a flow transducer or transmitter has to be installed in line whereas for measuring pressure, pressure transmitters can be mounted independently with a tubing connection to the pipe, otherwise known as remote monitoring.

The basic fault finding tool in any pneumatic or hydraulic system is the pressure gage. An example of a test pressure gage which measures gage pressure is the simple Bourdon pressure gage. A Bourdon pressure gage consists of a flattened 'C' shaped tube, which is fixed at one end. When pressure is applied to the tube, it tends to straighten, with the free end moving up and to the right. For low-pressure ranges, a spiral tube is used to increase its sensitivity.

The movement of the tube is converted into a circular pointer movement by a mechanical quadrant and pinion. The construction of a simple bourdon pressure gage is shown in Figure 2.3(a).

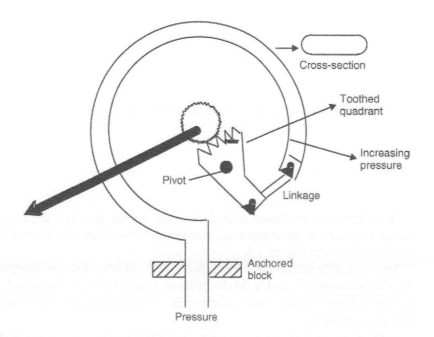

Figure 2.3(a)
A simple bourdon pressure gage

If an electrical output signal is required for a remote indication, the pointer can be replaced by a potentiometer as shown in Figure 2.3(b).

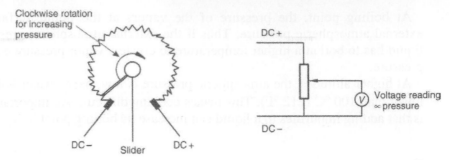

Figure 2.3(b)
An electrical signal from the bourdon gage

Hydraulic and pneumatic systems tend to exhibit large pressure spikes as the load accelerates and decelerates. These spikes can be misleading especially with regard to the true value measured and also end up causing damage to the pressure gage. In order to avoid this, a snubber restriction is provided to dampen the response of a pressure sensor. This has been illustrated in Figure 2.3(c).

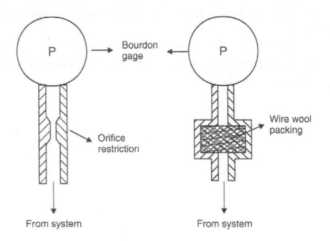

Figure 2.3(c)
Snubber restrictions

Bourdon pressure-based transducers are robust but low-accuracy devices. For more accurate pressure measurement, transducers based on the forced balance principle are used as shown in Figure 2.4.

This is a differential pressure transducer in which a low-pressure inlet (LP) is left open to the atmosphere and a high-pressure inlet (HP) is connected to the system. The difference between the two readings (HP – LP) obtained in the form of a signal, indicates the gage pressure.

A pressure increase in the system, deflects the pressure sensitive diaphragm to the left. This movement is detected by the transducer and which, through a servo amplifier, leads to an increase in the coil current. The current through the transducer is proportional to the differential pressure as the force from the balance coil exactly balances the force arising from the differential pressure between the LP and HP. Pressure does not depend on the shape or size of the container.

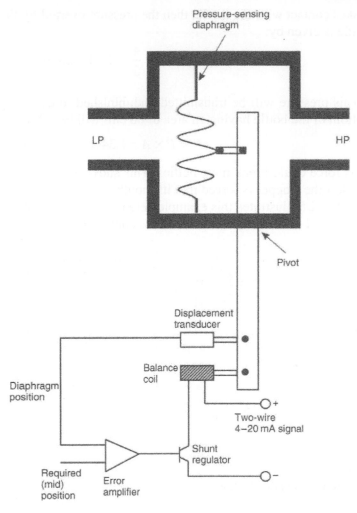

Figure 2.4
Forced balanced pressure transducer

2.3 Pascal's law

The underlying principle of how fluids transmit power is revealed by Pascal's law. Pascal's law states that the pressure applied to a confined fluid is transmitted undiminished in all directions. This law forms the basis for understanding the relationship between force, pressure and area, which can be mathematically expressed as:

$$\text{Force} = \text{Pressure} \times \text{Area} \quad \text{or}$$

$$\text{Pressure} = \frac{\text{Force}}{\text{Area}}$$

The transmitted pressure acts with equal force on every unit area of the containing vessel and in a direction at right angle to the surface of the vessel exposed to the liquid.

Pascal's law can be illustrated by the following example.

A bottle is filled with a liquid, which is not compressible. A force of 4 kg is applied to the stopper whose surface area is 3 cm^2. Let's assume that the area of the bottle bottom is 60 cm^2. If the stopper is inserted into the bottle mouth, with a force of 4 kg such that it makes contact with the liquid, then the pressure exerted by the stopper on the liquid in the bottle is given by:

$$P = \frac{4}{3} = 1.34 \, \text{kgf/cm}^2$$

This pressure will be transmitted undiminished to every square area of the bottle. The bottom of the bottle having an area of 60 cm² will be subjected to an additional force of:

$$F = P \times A = 1.34 \times 60 = 80.4 \, \text{kgf}$$

This force could break most bottles. This shows why a glass bottle filled with liquid can break if the stopper is forced into its mouth.

Figure 2.5 illustrates this example better. It also substantiates the fact that pressure does not depend on the shape and size of the container.

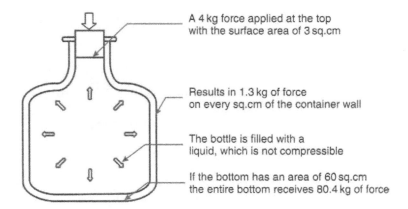

A 4 kg force applied at the top with the surface area of 3 sq.cm

Results in 1.3 kg of force on every sq.cm of the container wall

The bottle is filled with a liquid, which is not compressible

If the bottom has an area of 60 sq.cm the entire bottom receives 80.4 kg of force

Figure 2.5
Demonstration of Pascal's law

2.4   Application of Pascal's law

In this section, we shall study two basic applications of Pascal's law, the hydraulic jack and the air-to-hydraulic booster.

2.4.1   Hydraulic jack

This system uses a piston-type hand pump to power a single acting hydraulic cylinder as illustrated in Figure 2.6.

A hand force is applied at point 'A' of handle 'ABC', which pivots about point 'C'. The piston rod of the hand pump is pinned to the input handle at point 'B'. The hand pump contains a cylinder for aiding the up and down movement. When the handle is pulled, the piston moves up, thereby creating a vacuum in the space below it. As a result of this, the atmospheric pressure forces the oil to leave the oil tank and flow through check valve 1. This is the suction process.

When the handle is pushed down, oil is ejected from the hand pump and flows through the check valve 2. Oil now enters the bottom of the load cylinder. The load cylinder is similar in construction to the pump cylinder. Pressure builds up below the load piston as oil is ejected from the pump. From Pascal's law, we know that the pressure acting on the load piston is equal to the pressure developed by the pump below its piston. Thus each time the handle is operated up and down, a specific volume of oil is ejected from the pump to lift the load cylinder to a given distance against its load resistance. The bleed valve is a hand-operated valve which when opened, allows the load to be lowered by bleeding oil from the load cylinder back to the oil tank. This cylinder is referred to as single acting because it is hydraulically powered in one direction only.

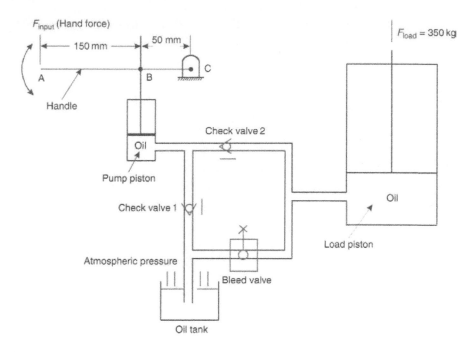

Figure 2.6
Hand-operated hydraulic jack system

2.4.2 Air-to-hydraulic pressure booster

Air-to-hydraulic pressure booster is a device used to convert workshop air into a higher hydraulic pressure needed for operating cylinders requiring small to medium volumes of high-pressure oil (Figure 2.7(a)).

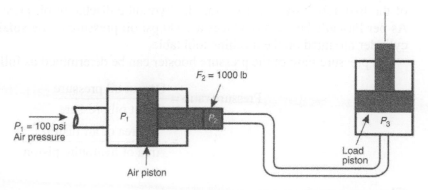

Figure 2.7(a)
An air-to-hydraulic system

It consists of an air cylinder with a large diameter driving a small diameter hydraulic cylinder. Any workshop equipped with an airline can easily obtain hydraulic power from an air-to-hydraulic booster hooked into the airline. Figure 2.7(b) shows an application of the air-to-hydraulic booster. Here the booster is seen supplying high-pressure oil to a hydraulic cylinder used to clamp a work piece to a machine tool table.

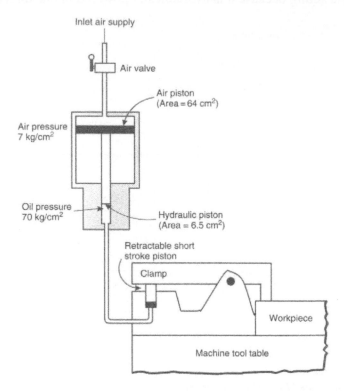

Figure 2.7(b)
Manufacturing application of an air-to-hydraulic booster

Since the workshop air pressure normally operates at around 100 psi, a pneumatically operated clamp would require a relatively larger cylinder to hold the work piece while it is being machined.

Let us assume that the air piston has a 10 sq. in. area and subjected to a pressure of 100 psi. This produces a 1000 lb force on the hydraulic cylinder piston. Thus if the area of the hydraulic piston is 1 sq. in., the hydraulic discharge oil pressure will be 1000 psi. As per Pascal's law this produces a 1000 psi oil pressure at the small hydraulic clamping cylinder mounted on the machine tool table.

The pressure ratio of the pressure booster can be determined as follows:

$$\text{Pressure ratio} = \frac{\text{Output oil pressure}}{\text{Input oil pressure}}$$
$$= \frac{\text{Area of air piston}}{\text{Area of hydraulic piston}}$$

2.5   Flow

Pascal's law holds good only for liquids, which are at rest or in the static state. As stated earlier, the study of this science dealing with liquids at rest is referred to as Hydrostatics. The study of liquids in motion can be discussed under two headings, Hydrokinetics and Hydrodynamics.

Hydrokinetics deals with the motion of fluid particles without considering the forces causing the motion. The velocity at any point in the flow field at any time is studied in this branch of fluid mechanics. Once this velocity is known, the pressure distribution and the forces acting on the fluid can be determined. Hydrodynamics is the study of fluid motion that includes the forces causing the flow.

Fluid motion can be described by two methods. They are:

1. Lagrangian method
2. Eulerian method.

In the Lagrangian method a single fluid particle is followed during its motion and its characteristics such as pressure, density, velocity, acceleration, etc. are described.

In the Eulerian method, any point in the space occupied by the fluid is selected and an observation is made on the changes in parameters such as pressure, density, velocity, and acceleration at this point. The Eulerian method is generally followed and is most preferred, when it comes to analyzing hydraulic systems.

No study of flow is complete without understanding three important principles related to the phenomenon of flow, which are as follows.

1. *Flow makes it go*: The actuator must be supplied with flow for anything in a hydraulic system to move. The cylinder is normally retracted and requires flow to extend itself. The extension and retraction functions are accomplished with the help of a direction control valve.
2. Rate of flow determines speed: The rate of flow usually measured in gallons per minute or gpm is determined by the pump. The speed of the actuator changes with variation in pump outlet flow.
3. Changes in actuator volume displacement will change actuator speed at a given flow rate: When the cylinder retracts, less volume needs to be displaced because of the space occupied by the cylinder rod. This results in a faster actuator cycle. Therefore, there is always a difference in actuator speed between the extend and retract functions.

2.5.1 Meaning of flow

Flow velocity is very important in the design of a hydraulic system. When we speak of fluid flow down a pipe in a hydraulic system, the term flow in itself conveys three distinct meanings, which are:

1. *Volumetric flow*, which is a measure of the volume of a fluid passing through a point in unit time.
2. *Mass flow*, which is a measure of the mass of a fluid passing through a point in unit time.
3. *Velocity of flow*, which is a measure of the linear speed of a fluid passing through the point of measurement.

2.5.2 Types of fluid flow

Fluid flow can be classified as follows:

- Steady and unsteady flows
- Uniform and non-uniform flows
- Laminar and turbulent flows
- Rotational and non-rotational flows.

Steady flow

Fluid flow is said to be steady if at any point in the flowing fluid, important characteristics such as pressure, density, velocity, temperature, etc. that are used to describe the behavior of a fluid, do not change with time. In other words, the rate of flow through any cross-section of a pipe in a steady flow is constant.

Unsteady flow

Fluid flow is said to be unsteady if at any point in the flowing fluid any one or all the characteristics describing the behavior of a fluid such as pressure, density, velocity and temperature change with time. Unsteady flow is that type of flow, in which the fluid characteristics change with respect to time or in other words, the rate of flow through any cross-section of a pipe is not constant.

Uniform flow

Flow is said to be uniform, when the velocity of flow does not change either in magnitude or in direction at any point in a flowing fluid, for a given time. For example, the flow of liquids under pressure through long pipelines with a constant diameter is called uniform flow.

Non-uniform flow

Flow is said to be non-uniform, when there is a change in velocity of the flow at different points in a flowing fluid, for a given time. For example, the flow of liquids under pressure through long pipelines of varying diameter is referred to as non-uniform flow.

All these type of flows can exist independently of each other. So there can be any of the four combinations of flows possible:

1. Steady uniform flow
2. Steady non-uniform flow
3. Unsteady uniform flow
4. Unsteady non-uniform flow.

Laminar flow

A flow is said to be laminar if the fluid particles move in layers such that one layer of the fluid slides smoothly over an adjacent layer. The viscosity property of the fluid plays a significant role in the development of a laminar flow. The flow pattern exhibited by a highly viscous fluid may in general be treated as laminar flow (Figure 2.8(a)).

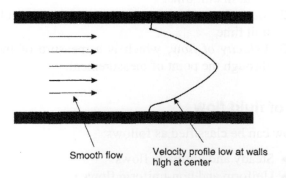

Smooth flow   Velocity profile low at walls
high at center

Figure 2.8(a)
Laminar flow

Turbulent flow

If the velocity of flow increases beyond a certain value, the flow becomes *turbulent*. As shown in Figure 2.8(b), the movement of fluid particles in a turbulent flow will be random. This mixing action of the colliding fluid particles generates turbulence, thereby resulting in more resistance to fluid flow and hence greater energy losses as compared to laminar flow.

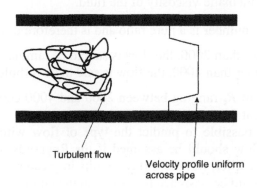

Turbulent flow

Velocity profile uniform
across pipe

Figure 2.8(b)
Turbulent flow

The frictional resistance that a fluid moving in a pipe encounters, is normally proportional to the velocity of flow. However, once the flow turns turbulent, this frictional resistance encountered by the liquid becomes proportional to the square of the velocity of fluid flow.

$$F \, \alpha \, v \quad \text{for laminar flow}$$
$$F \, \alpha \, v^2 \quad \text{for turbulent flow}$$

Where
    'F' is the resistance to fluid flow
    'v' is the velocity of flow.

Due to greater energy losses, turbulent flow is generally avoided in *hydraulic systems*. Some of the causes for turbulent flow in a hydraulic system are:

- Roughness of pipelines
- Obstructions to flow
- Degree of curvature of bends
- Increase in the number of bends.

Reynolds number

In a hydraulic system, it is important to know whether the flow pattern inside a pipe is laminar or turbulent and also to determine the conditions that govern the transition of the flow from laminar to turbulent. This is where Reynolds number holds much significance. The experiments performed by Osborn Reynolds led to important conclusions through which the nature of flow could be determined, by using a parameter known as the 'Reynolds number'.

Reynolds number 'R_e' is given by the expression:

$$R_e = \frac{vd}{\eta}$$

Where

v is the velocity of flow

d is the diameter of the pipe

η is the kinematic viscosity of the fluid.

Reynolds number is a pure ratio and is therefore dimensionless.

If R_e is lesser than 2000, the flow is said to be laminar

If R_e is greater than 4000, the flow is said to be turbulent.

Any value of R_e ranging between 2000 and 4000 covers a critical zone between laminar and turbulent flow.

It is not possible to predict the type of flow within the critical zone. But normally, turbulent flow should be assumed if the Reynolds number lies in the critical zone. As mentioned earlier, turbulent flow results in greater energy losses and therefore hydraulic systems should be designed to operate in the laminar flow region.

The greater energy losses that arise as a consequence of turbulent flow result in an increase in the temperature of the fluid. This condition can be alleviated to a great extent by providing for a slight increase in the pipe size in order to establish laminar flow.

Rotational flow

A flow is said to be rotational if the fluid particles moving in the direction of flow rotate around their own axis.

Non-rotational flow

If the fluid particles flowing in a laminar pattern do not rotate about their axis, then the flow is said to be non-rotational.

2.5.3   Rate of flow or discharge (*Q*)

The rate of flow or discharge is defined as the quantity of fluid flowing per second, through a pipe or channel section. In the case of incompressible fluids (liquids), the discharge is expressed in terms of the volume of fluid flowing across the section per second.

$$\text{Flow rate (liquid)} = \frac{\text{Volume}}{\text{Time}}$$

For compressible fluids (gases) the discharge is expressed as the weight of the fluid flowing across a section per second. So obviously the units of flow rate or discharge (Q) are: m^3/s or liters/s for liquids and kgf/s or N/s for gases.

Consider a liquid flowing through a pipe of cross-sectional area 'A' and an average flow velocity 'v' across the section. We then have

$$\text{Flow rate or discharge } Q = \frac{\text{Volume}}{\text{Time}}$$

$$= \frac{(\text{Area} \times \text{Distance})}{\text{Time}}$$

This can be mathematically represented as

$$Q = A \times v \left[\frac{\text{Since distance}}{\text{Time}} = \text{Velocity}\,(v) \right]$$

2.5.4 Law of conservation of energy

As discussed earlier, the law of conservation of energy states that energy can neither be created nor destroyed, but can be transformed from one form to the other. This also means that the total energy of the system at any location remains constant.

The total energy of a liquid in motion includes:

- Potential energy
- Kinetic energy and
- Internal energy.

Potential energy (PE)

It is the energy stored in the system due to its position in the gravitational force field. If a heavy object such as building stone is lifted from the ground to the roof, the energy required to lift the stone is stored in it as potential energy. This stored potential energy remains unchanged as long as the stone remains in position.

Potential energy can be mathematically represented as:

$$PE = Z \times g$$

Where
   Z is the height of the object above the datum and
   g is the acceleration due to gravity.

Kinetic energy (KE)

This is the energy possessed by the system by virtue of its motion and is given by the equation.

$$KE = \frac{W}{2gv^2}$$

Where
   W is the weight of the system under consideration
   g is the acceleration due to gravity and
   v the velocity of the system.

To quote an example, if a body weighing 1 kg is moving with a velocity of v m/s with respect to the observer, then the kinetic energy stored in the body is given by:

$$KE = \frac{v^2}{2}$$

This energy will remain stored in the body as long as it continues to be in motion at a constant velocity. When the velocity is zero, the kinetic energy is also zero.

Internal energy

Molecules possess mass. They also possess motion which is translational and rotational in nature, in both the liquid as well as the gaseous states. Owing to this mass and motion, these molecules have a large amount of kinetic energy stored in them. Any change in temperature results in a change in the molecular kinetic energy since molecular velocity is a function of temperature.

Also the molecules are attracted towards each other by very large forces in their solid state. These forces tend to vanish once a perfect gas state is reached. During the melting process of a solid or the vaporization process of a liquid, it is necessary to overcome these forces. The energy required to bring about this change is stored in the molecules as potential energy.

The sum of these energies is called the internal energy, which is stored within the body. We refer to this energy as internal energy or thermal energy and it is denoted by the symbol 'u'.

2.5.5 Bernoulli's equation

An important equation formulated by an eighteenth-century Swiss scientist Daniel Bernoulli and known as Bernoulli's equation is one of the vital tools employed in the analysis of hydraulic systems. By applying this principle in the design of a hydraulic system, it is possible to size various components comprising the system such as pumps, valves and piping, for effective and proper system operation.

Bernoulli's equation basically enunciating the principle of conservation of energy states that in a liquid flowing continuously, the sum total of static, pressure and velocity energy heads is constant at all sections of the flow. The law as applied to a hydraulic pipeline is illustrated in Figure 2.9.

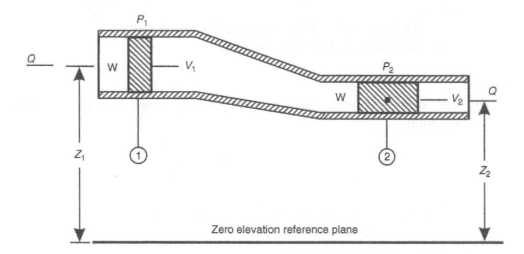

Figure 2.9
Pipeline for deriving Bernoulli's equation

In the above figure, consider a fluid flowing through a hydraulic pipeline at section 1 where

W is the weight of the fluid
Z_1 is the elevation at which the fluid is flowing
V_1 is the velocity of the fluid
P_1 is the pressure exerted by the fluid.

When this fluid arrives at section 2, assume that its elevation is 'Z_2', velocity is V_2, and pressure is P_2.

According to Bernoulli's principle, total energy possessed by the fluid at section 1 = Total energy possessed by the fluid at section 2

$$WZ_1 + W\frac{P_1}{\rho} + \left(\frac{W}{2g}\right)v_1^2 = WZ_2 + W\frac{P_2}{\rho} + \left(\frac{W}{2g}\right)v_2^2$$

Since liquids are considered to be incompressible the density is the same and therefore the equation for a fluid of unit weight reduces to:

$$Z_1 + \frac{P_1}{\rho} + \frac{v_1^2}{2g} = Z_2 + \frac{P_2}{\rho} + \frac{v_2^2}{2g}$$

The use of the expression 'head' has gained widespread acceptance and accordingly

Z is called the *elevation* or *potential head*
P/ρ is called the *pressure head* and
$v^2/2g$ is called the *velocity head.*

Further corrections to the above equation can be made by taking into account the following factors:

1. The frictional resistance to motion when the fluid passes through the pipe from section 1 to section 2 in overcoming which, a part of the fluid energy is lost. Let h_f represent the *energy head* lost due to friction in the pipeline.
2. Assuming the presence of a pump and a motor between sections 1 and 2.

Let h_p known as *pump head* represent the energy per unit weight of the fluid added by the pump and h_m known as *motor head* represent the energy per unit weight utilized or removed by the motor. This leads us to the corrected Bernoulli's equation which is

$$Z_1 + \frac{P_1}{\rho} + \frac{v_1^2}{2g} + h_p - h_f - h_m = Z_2 + \frac{P_2}{\rho} + \frac{v_2^2}{2g}$$

We shall now discuss the means by which the magnitude of the head loss can be evaluated.

The total head loss in the system can be further categorized as:

- Losses occurring in pipes and
- Losses occurring in fittings.

Head losses due to friction in pipes can be found by using the *Darcy's equation*, which is

$$h_f = \frac{f\,Lv^2}{2\,gd}$$

Where
  f is Darcy's frictional coefficient or factor
  L is the length of the pipe
  v is the average fluid velocity
  d is the inside pipe diameter
  g is the acceleration due to gravity.

Darcy's equation can be applied for calculating the head loss due to friction, for both laminar as well as turbulent flows. The only difference will be in the evaluation of the frictional coefficient 'f'.

Frictional losses in laminar flow

For laminar flow, the friction factor 'f' is given by

$$f = \frac{64}{R_e}$$

Where R_e is the Reynolds number.

Substituting for $f = 64/R_e$ in the above equation, we have

$$h_f = \frac{64 \, Lv^2}{R_e \, 2gd}$$

which is called the Hagen-Poiseuille equation.

Frictional losses in turbulent flow

Unlike in the case of laminar flow, the friction factor cannot be represented by a simple formula for turbulent flow. This is due to the fact that the movement of fluid particles in a turbulent flow is random and fluctuating in nature. Here the friction factor has been found to depend not only on the Reynolds number but also the relative roughness of the pipe. This relative roughness is given by:

$$\text{Relative roughness} = \frac{\text{Pipe inside surface roughness}}{\text{Pipe inside diameter}} = \frac{\epsilon}{D}$$

Figure 2.10 illustrates the physical meaning of the pipe inside surface roughness ϵ, called the absolute roughness.

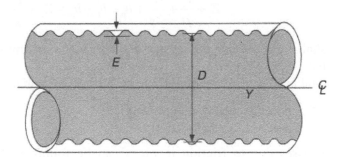

Figure 2.10
Absolute roughness of a pipe

The absolute roughness depends on the pipe material as well as the method of manufacture. Another point to be noted is that the roughness values of pipes undergo significant changes over a period of time due to deposit buildup on the walls.

2.5.6   Pressure–flow relationship

Figure 2.11 shows a venturi essentially consisting of the following three sections:

      1. Converging part
      2. Throat and
      3. Diverging part.

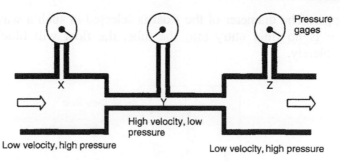

Figure 2.11
Relationship between flow and pressure

Let 'v_1' and 'v_2' be the velocities of the fluid at the converging part which is section 1 and the throat which is section 2, respectively. From the continuity equation, we know that the flow velocity at the throat 'v_2' is greater than 'v_1'. Bernoulli's equation for the flow between sections 1 and 2 can be written as

$$\frac{P_1}{\rho} + \frac{v_{12}}{2g} = \frac{P_2}{\rho} + \frac{v_{22}}{2g}$$

Since the pipe is horizontal, potential energy at both the sections is constant. Note that whenever the flow lines are located with a small difference in level, the potential head can be neglected.

Now, since v_2 is greater than v_1, the pressure P_1 must be greater than P_2. This is in accordance with the law of conservation of energy, because if the fluid has gained kinetic energy by passing from section 1 to section 2, then it has to lose pressure energy in order to conform to the law of conservation of energy. Again, at the diverging portion of the pipe, the pressure recovers and the flow velocity falls.

2.6   Flow measurement

In order to troubleshoot hydraulic systems and to evaluate the performance of hydraulic components, it is often required to measure the flow rate. For example, flow measurements are undertaken to check the volumetric efficiency of pumps and also to determine leakage paths within a hydraulic system.

Although, there are numerous flow measuring devices for measuring flow in a hydraulic circuit, our discussion is limited to the three most commonly employed, which are:

1. Rotameter
2. Turbine flowmeter and
3. Orifice plate flowmeter.

2.6.1   Rotameter

The Rotameter also known as variable area flowmeter is the most common among all flow measurement devices. Figure 2.12 shows the operation of a Rotameter. It basically consists of a tapered glass tube calibrated with a metering float that can move vertically up and down in the glass tube. Two stoppers one at the top and the other at the bottom of the tube prevent the float from leaving the glass tube. The fluid enters the tube through the inlet provided at the bottom. When no fluid is entering the tube, the float rests at the bottom of the tapered tube with one end of the float making contact with the lower stopper. The diameter of the float is selected in such a way that under conditions where there is no fluid entry into the tube, the float will block the small end of the tube completely.

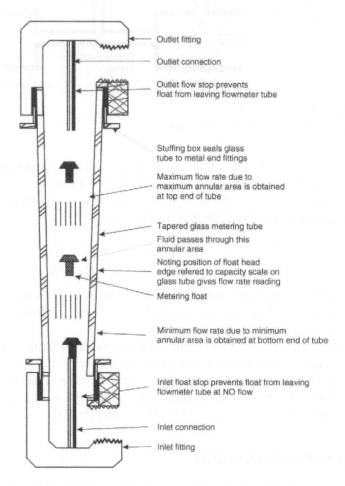

Figure 2.12
Operation of a Rotameter (Courtesy of Fischer & Porter Company, Pennsylvania)

When the fluid starts entering the tube through the inlet provided at the bottom, it forces the float to move upwards. This upward movement of the float will continue, until an equilibrium position is reached at which point the weight of the float is balanced by the upward force exerted by the fluid on the float. Greater the flow rate, higher is the float rise in the tube. The graduated tube allows direct reading of the flow rate.

2.6.2   Turbine-type flowmeter

Figure 2.13 is a simple illustration of a turbine-type flowmeter.

This flowmeter has a turbine rotor in the housing, which is connected to the pipeline whose flow rate is to be measured. When the fluid flows, it causes the turbine to rotate. Higher the flow rate, greater is the speed of the turbine. The magnetic end of a sensor, which is positioned near the turbine blades, produces a magnetic field whose magnetic lines of force are interrupted by the rotation of the turbine blades, thereby generating an electrical impulse. An electrical device connected to the sensor converts the pulses to flow rate information.

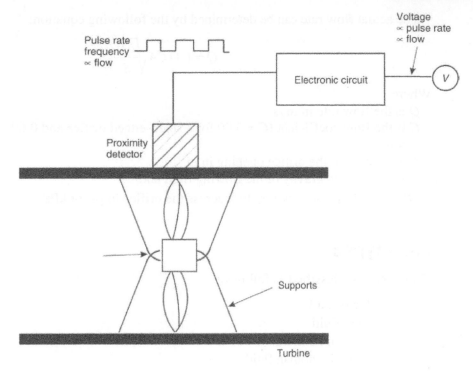

Figure 2.13
Turbine flowmeter

2.6.3 Orifice plate-type flowmeter

Another method by which flow rate can be determined involves the use of an orifice plate-type flowmeter in which an orifice is installed in the pipeline as shown in Figure 2.14.

The figure also shows the presence of two pressure gages, one each on either side of the orifice. This arrangement enables us to determine the pressure drop (ΔP) across the orifice when the fluid flows through the pipe and given by $\Delta P = P_1 - P_2$. The higher the flow rate, greater will be the pressure drop.

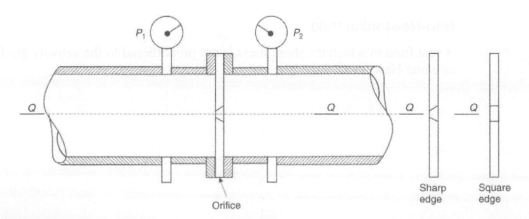

Figure 2.14
Orifice-type flowmeter

The actual flow rate can be determined by the following equation:

$$Q = 1.41CA\sqrt{\frac{\Delta P}{S}}$$

Where

Q is the flow rate in m^3/s

C is the flow coefficient (C = 0.80 for a sharp-edged orifice and 0.60 for a square-edged orifice)

A is the area of the orifice opening in m^2

S is the specific gravity of the flowing fluid and

$\Delta P = P_1 - P_2$ is the pressure drop across the orifice in psi or kPa.

2.7　Fluid types

Fluids may be classified as follows:

- Ideal fluid
- Real fluid
- Newtonian fluid
- Non-Newtonian fluid.

Ideal fluid

An ideal fluid is one, which is incompressible and has no viscosity. Such a fluid is only imaginary, as all existing fluids possess some viscosity.

Real fluid

Simply speaking, a fluid which possesses viscosity is known as a real fluid. All fluids, in actual practice are real fluids.

Newtonian fluid

During the course of our discussion on viscosity, we have seen that shear stress is proportional to the velocity gradient i.e. $\tau \alpha$ dv/dy. A real fluid in which the shear stress is proportional to the velocity gradient is known as a Newtonian fluid.

Non-Newtonian fluid

A real fluid in which the shear stress is not proportional to the velocity gradient is known as a non-Newtonian fluid.

3

Hydraulic pumps

3.1 Objectives

After reading this chapter, the student will be able to:

- Distinguish between positive and non-positive displacement pumps
- Understand the principle of operation of gear, vane and piston pumps
- Differentiate between fixed and variable displacement pumps, external and internal gear pumps as well as axial and radial piston pumps
- Explain how pressure-compensated pumps work
- Identify the various types of pumps used in hydraulics
- Select and size pumps for various hydraulic applications
- Carry out basic maintenance activities on the pumps.

3.2 Principle of operation

The sole purpose of a pump in a hydraulic system is to provide flow. A pump, which is the heart of a hydraulic system, converts mechanical energy, which is primarily rotational power from an electric motor or engine, into hydraulic energy. While mechanical rotational power is the product of torque and speed, hydraulic power is pressure times flow. The pump can be designed in such a way that either flow or pressure is fixed, while the other parameter is allowed to swing with the load. In other words, by fixing the pump flow, the pressure goes up as the load restriction is increased. Conversely, the flow goes down with an increase in load restriction when the pump delivers fixed pressure.

The pumping action is the same for every pump. Due to mechanical action, the pump creates a partial vacuum at the inlet. This causes the atmospheric pressure to force the fluid into the inlet of the pump. The pump then pushes the fluid into the hydraulic system (Figure 3.1).

The pump contains two check valves. Check valve 1 is connected to the pump inlet and allows fluid to enter the pump only through it. Check valve 2 is connected to the pump discharge and allows fluid to exit only through it.

When the piston is pulled to the left, a partial vacuum is created in the pump cavity 3. This vacuum holds the check valve 2 against its seat and allows atmospheric pressure to push the fluid inside the cylinder through the check valve 1. When the piston is pushed to the right, the fluid movement closes check valve 1 and opens outlet valve 2. The quantity of fluid displaced by the piston is forcibly ejected from the cylinder. The volume of the fluid displaced by the piston during the discharge stroke is called the displacement volume of the pump.

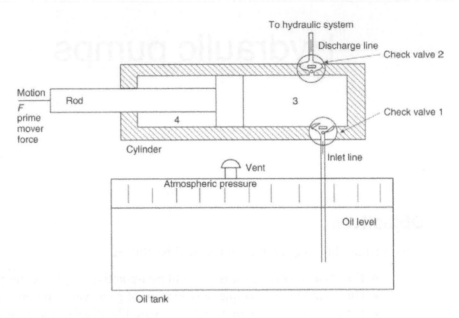

Figure 3.1
Pumping action of a simple piston pump

3.3 Pump classification

Pumps can be broadly listed under two categories:

1. Non-positive displacement pumps and
2. Positive displacement pumps.

3.3.1 Non-positive displacement pumps

They are also known as hydro-dynamic pumps. In these pumps the pressure produced, is proportional to the rotor speed. In other words, the fluid is displaced and transferred using the inertia of the fluid in motion. These pumps are incapable of withstanding high pressures and are generally used for low-pressure and high-volume flow applications. Normally their maximum pressure capacity is limited to 20–30 kgf/cm^2. They are primarily used for transporting fluids from one location to the other and find little use in the hydraulic or fluid power industry.

Because of fewer numbers of moving parts, non-positive displacement pumps cost less and operate with little maintenance. They make use of Newton's first law of motion to move the fluid against the system resistance. Although these pumps provide a smooth and continuous flow, their flow output is reduced as the system resistance (resistance to flow) is increased. In fact it is possible to completely block the outlet to stop all flow even while the pump is running at the designed speed. Thus the pump flow rate depends not only on the rotational speed (rpm) at which it is driven but also on the resistance of the external system. As the resistance of the external system increases, some of the fluid will slip back, causing a reduction in the discharge flow rate. When the resistance of the external system becomes very large, the pump will produce no flow and thus its volumetric efficiency becomes zero. Examples of these pumps are the centrifugal and axial (propeller) pumps.

In a centrifugal pump, a simple sketch of which is illustrated in Figure 3.2, rotational inertia is imparted to the fluid. Centrifugal pumps are not self-priming and must be positioned below the fluid level.

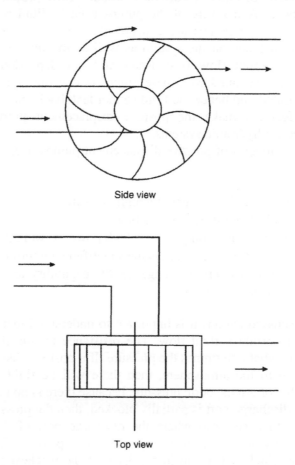

Side view

Top view

Figure 3.2
A typical centrifugal pump

Principle of operation

The fluid from the inlet port enters at the center of the impeller. The rotating impeller imparts centrifugal force to the fluid and causes it to move radially outward. This results in the fluid being forced through the outlet discharge port of the housing. The tips of the impeller blades merely move through the fluid while the rotational speed maintains the fluid pressure corresponding to the centrifugal force established.

Centrifugal pumps are generally used in pumping stations, for delivering water to homes and factories. The advantages of non-positive displacement pumps are:

- Low initial cost and minimum maintenance
- Simplicity of operation and high reliability
- Capable of handling any type of fluid, for example sludge and slurries.

Since the impeller imparts kinetic energy to the fluid, centrifugal pumps are also known as hydrokinetic power generators.

3.3.2   Positive displacement or hydrostatic pumps

As the name implies, these pumps discharge a fixed quantity of oil per revolution of the pump shaft. In other words, they produce flow proportional to their displacement and rotor speed. A majority of the pumps used in fluid power applications belong to this category. These pumps are capable of overcoming the pressure that results from the mechanical loads on the system as well as the resistance to flow due to friction. Thus the pump output flow is constant and not dependent on system pressure. Another advantage associated with these pumps is that the high-pressure and low-pressure areas are separated and hence the fluid cannot leak back and return to the low-pressure source. These features make the positive displacement pump most suited and universally accepted for hydraulic systems.

The advantages of positive displacement pumps over non-positive displacement pumps are:

- Capability to generate high pressures
- High volumetric efficiency
- Small and compact with high power to weight ratio
- Relatively smaller changes in efficiency throughout the pressure range
- Wider operating range i.e. the capability to operate over a wide pressure and speed range.

As discussed earlier, it is important to understand that pumps do not produce pressure; they only produce fluid flow. The resistance to this flow as developed in a hydraulic system is what determines the pressure. If a positive displacement pump has its discharge port open to the atmosphere, then there will be fluid flow, but no discharge pressure above that of atmospheric pressure, because there is no resistance to flow.

If the discharge port is partially blocked, then the pressure will rise due to the resistance to flow. In a scenario where the discharge port of the pump is completely blocked, theoretically an infinite resistance to flow is possible. This will result in a rapid rise in pressure which will result in breakage of the weakest component in the circuit. This is exactly the reason why positive displacement pumps are provided with safety controls, which help prevent the rise in pressure beyond a certain value.

A detailed classification of pumps is shown in Figure 3.3.

3.4   Gear pump

Gear pumps as the name suggests make use of the principle of two gears in mesh in order to generate pumping action. They are compact, relatively inexpensive and have few moving parts. Gear pumps are further classified as:

- External gear pumps
- Internal gear pumps
- Lobe pumps and
- Gerotor pumps.

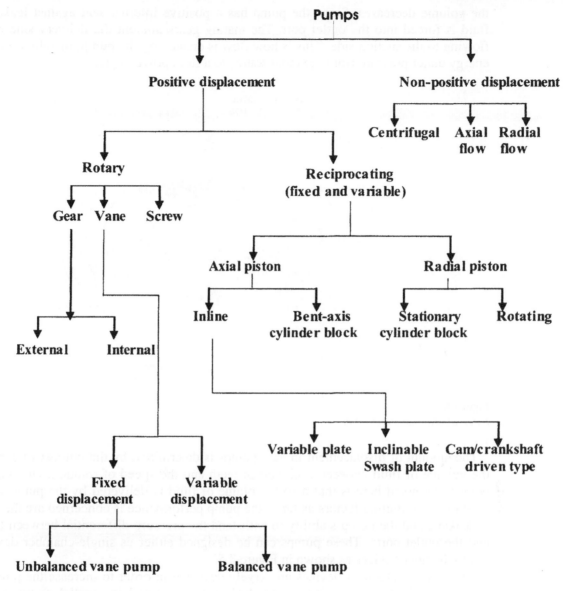

Figure 3.3
Classification of pumps

3.4.1 External gear pump

A schematic of an external gear pump is shown in Figure 3.4.

An external gear pump consists of two gears usually equal in size, which mesh externally and are housed in a pump case. Each gear is mounted on a shaft, which is supported by needle bearings in the case covers. One of these shafts is coupled to a prime mover and is called the drive shaft. The gear mounted on this shaft is called the drive gear. It drives the second gear as it rotates. Two side plates are provided one on each side of the gear. The side plates are held between the gear case and case covers.

The suction side is where the teeth come out of the mesh. The volume expands in this region leading to a drop in pressure below the atmospheric value, which results in the fluid getting pushed into this void. The fluid is trapped between the housing and the rotating teeth of the gears. The discharge side is where the teeth go into the mesh and here the volume decreases. Since the pump has a positive internal seal against leakage, the fluid is forced into the outlet port. The mating gears prevent the delivery side oil from flowing to the suction side. This is how flow is created by the pump, thereby transferring energy under pressure from the input source to a fluid power actuator.

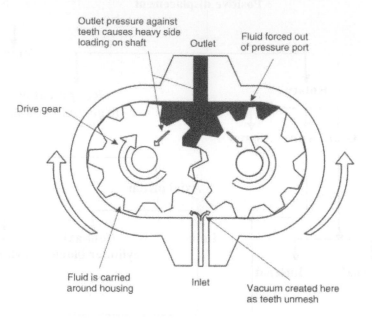

Figure 3.4
External gear pump operation

The amount of displacement in gear pumps is determined by the number of gear teeth, the volume of fluid between each pair of teeth and the speed of rotation. One important point to be noted here is that a fixed volume of fluid is delivered by the pump for each rotation. The limiting factors as far as the pump performance is concerned are the amount of leakage and the pump's ability to withstand the pressure differential between the inlet and the outlet ports. These pumps can be designed either as single-chamber devices or multi-chamber devices as shown in Figure 3.5.

The multiple chamber devices are largely designed in order to increase the pump flow capacity, and consists of either individual units connected in parallel or by providing many pumps on a single shaft by isolating the inlets and the outlets. One important drawback with external gear pumps is the unbalanced side load on its bearings, caused due to high pressure at the outlet and low pressure at the inlet which in turn results in slower speeds and lower pressure ratings in addition to reducing the bearing life.

It is important for a gear pump to have the following:

- Close meshing of gears
- A very small clearance between the gear teeth and housing and also between the side plates and gear face.

Two mating teeth coming into mesh during normal operation of a gear pump will trap oil in the root section. Should there be no way for this oil to escape, an extremely high pressure would momentarily occur, thereby making the pump run with a great pounding or rattling noise. As the gear rotates, the 'trapping space' transmits from the discharge side to the suction side. The space gets smaller towards the center, becomes a minimum at the center and thereafter begins to increase.

Double pump

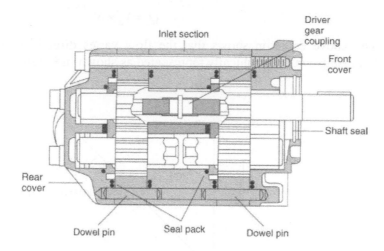

Figure 3.5
Combination of two gear pumps in parallel

In order to prevent the pounding or rattling of the pump, a relief recess is provided in the side plate, at a location corresponding to the trapping space in the meshing area. The recess extends only over that part of the trapping space. The pressurized oil is bled back to the discharge side through the recess. By the time the two teeth have arrived at their parting position, the trapping space will be way off the recess and so no internal communication occurs between the discharge and suction side through this recess.

During the course of overhauling/reconditioning of these pumps, a general tendency to carry out additional machining on the side plate in order to elongate the existing recess in excess of the specified length has been observed. This should be avoided, because too long a recess allows internal communication between the suction and discharge sides through it, even after the two teeth are parted from each other.

It is also equally important to ensure the correct positioning of the side plates at the time of pump build up, during overhaul. Should a side plate be placed in the reverse position, the recess comes into the position over the discharge side of the pump, thereby keeping the oil blocked in the trapping area.

The side plates in gear pumps are generally replaceable. The replacement is usually carried out during overhaul, if the clearance between the gear case and side plates has exceeded the allowable clearance limit specified by the manufacturer.

The following analysis evaluates the theoretical flow rate of a gear pump:
Let

D_o be the outside diameter of the gear teeth
D_i be the inside diameter of gear teeth
L be the width of gear teeth
V_d be the displacement volume of the pump
N be the speed of the pump and
Q_r be the theoretical pump flow rate.

From gear geometry, the volumetric displacement can be calculated from the equation:

$$V_d = \frac{\pi}{4}(D_o^2 - D_i^2)\, L$$

The theoretical flow rate is given by:

$$Q_t = V_d \times N$$

The above equation shows that the flow varies directly with the speed. This linear relationship can be graphically represented as shown in Figure 3.6(a).

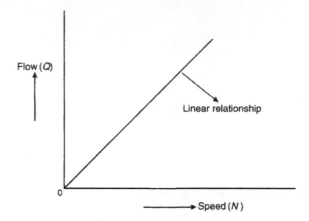

Figure 3.6(a)
Flow vs speed curve

A plot of flow vs pressure shows that the theoretical flow is constant at a given speed, as indicated by the solid line in the graph (Figure 3.6(b)).

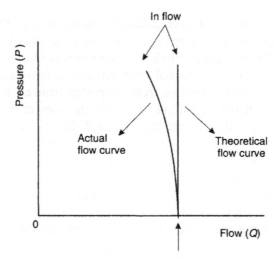

Figure 3.6(b)
Flow vs pressure curve at constant speed

However, as seen from the graph, the actual flow rate Q_a of the pump will be lower than the theoretical flow rate Q_t, as a result of internal leakage in the pump.

Internal leakage of oil

Oil from the high-pressure side usually tends to leak into the low-pressure side through any existing clearance. In a gear pump there is a running clearance between the gears and the case, which provides a path for leakage of oil. This results in a small amount of oil being continuously transferred to the low-pressure suction side from the high-pressure discharge side, in the form of internal leakage. This internal leakage of oil is known as *pump slippage*. This oil tends to flow along the shaft and then through the grooves provided on the back face of the side plate. From these grooves the oil enters the suction side through a drilled oil way provided in the pump case. It is this internal leakage oil that lubricates the needle bearings in the pump.

As a result of this leakage it is quite obvious that the actual flow rate of the gear pump will be less than the theoretical flow rate. The ratio of the actual flow rate and the theoretical flow rate is called the 'volumetric efficiency' of the pump which is represented by the symbol η_v

$$\eta_v = \left(\frac{Q_a}{Q_t} \right) \times 100$$

The higher the pump discharge pressure (resulting from a heavy load or resistance to flow in the hydraulic system) greater will be the internal leakage along with a correspondingly lower volumetric efficiency.

An abnormally high increase in pump discharge pressure apart from causing excessive internal leakage also results in a heavy load on the pump bearings. Consequently, the bearing life gets reduced while also causing all the gear teeth to scuff the bore wall, thereby resulting in damage to the pump.

As such, a gear pump or for that matter any positive displacement pump, requires protection from high pressures as discussed earlier. This is usually accomplished by incorporating a safety device called a 'relief valve'. Standard gear pumps are used at operating pressures and capacities up to 80 kg/cm^2 (1138 psi) and 700 liter/min (185 gpm) respectively, with peak pressure conditions varying between 120 (1707 psi) and 150 kg/cm^2 (2133 psi).

Gear pumps, which use spur gears (teeth parallel to the gear axis), are noisy in operation especially at high speeds. The use of helical gears (teeth inclined at a small angle to the gear axis) reduces the noise level and provides for smoother operation. However helical gears are expensive and are also limited to low-pressure applications. Another disadvantage with helical gears is the excessive end thrust that develops during the course of its operation. Herringbone gears are a good alternative especially in applications requiring higher pressures. They basically consist of two rows of helical teeth cut into one gear. One row in each gear is right-handed, while the other is left-handed. This is done to cancel out the axial thrust. A major plus point with these gears is the elimination of the end thrust. Additionally, they also provide smooth operation with a higher flow rate.

The pump case is a casting, with its bores machined in such a way as to provide a greater radial clearance on the suction side than on the discharge side in the quiescent state. This inequality is calculated to equalize the clearance when the pump is in its normal operating condition. In smaller-sized pumps the radial clearance on the discharge side is about 0.03 mm while in larger-sized pumps it is about 0.05 mm. The clearance on the suction side is about three times larger.

3.4.2   Internal gear pump

Internal gear pump is another variation of the basic gear pump. Figure 3.7 illustrates clearly, the internal construction and operation of an internal gear pump. The design consists of an internal gear, a regular spur gear, a crescent-shaped seal and an external housing. As power is applied to either gear, the motion of the gears draws the fluid from the suction, and forces it around both the sides of the crescent seal. This acts as a seal between the suction and discharge ports. When the teeth mesh on the side opposite to the crescent seal, the fluid is forced out through the discharge port of the pump. Similar to the external gear pump, internal gear pumps also have an in-built safety relief valve.

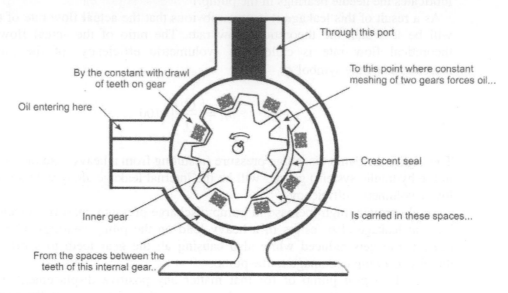

Figure 3.7
Operation of an internal gear pump

3.4.3 Lobe pump

The lobe pump is yet another variation of the basic gear pump. This pump operates in a fashion quite similar to that of an external gear pump, but unlike external gear pumps, the gears in these pumps are replaced with lobes which usually consist of three teeth. Figure 3.8 shows the operation of a lobe pump.

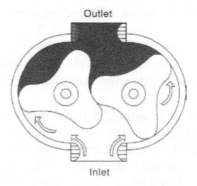

Figure 3.8
Operation of a lobe pump

Unlike the external gear pumps, both the lobes are driven externally so that they do not actually make contact with each other. They are quieter than the other gear pumps. Due to the smaller number of mating elements, the lobe pump will show a greater amount of pulsation. However, its volumetric displacement is generally greater than other types of gear pumps. Although these pumps have a low-pressure rating, they are well-suited for applications involving shear-sensitive fluids.

3.4.4 Gerotor pump

Gerotor pumps are one of the most common types of internal gear pumps whose operation is quite similar to that of an internal gear pump. The inner gear rotor (gerotor element) is power driven and draws the outer gear rotor around as they mesh together. This forms the inlet and outlet discharge pumping chambers between the rotor lobes. The tips of the inner and the outer lobes make contact to seal the pumping chambers from each other. The inner gear has one tooth less than the outer gear, and the volumetric displacement is determined by the space formed by the extra tooth in the outer gear.

Principle of operation

As the gear teeth pass the inlet, the volume capacity of the pumping chamber is expanded. This in turn results in the creation of a partial vacuum at the inlet, thereby allowing atmospheric pressure to push fluid into the pump. The fluid is then carried to the pump outlet between the inner and outer gear teeth which are in constant touch, thereby forming an effective seal. Like other gear pumps, the gerotor pumps also have a fixed displacement and an unbalanced bearing load and operate at lower capacities and pressures than most other pumps (Figure 3.9).

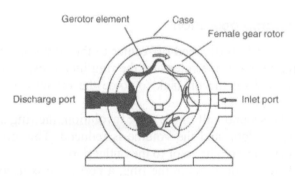

Figure 3.9
Operation of a gerotor pump

3.5 Vane pumps

Vane pumps have an inherent advantage as compared with gear pumps in that their basic design and construction minimizes the amount of leakage that occurs in gear pumps between the teeth gaps and also between the teeth and the pump housing. This is accomplished by using spring or hydraulically loaded vanes slotted into a driven rotor. Vane pumps generate a pumping action by causing the vanes to track along a ring. The pumping mechanism of a vane pump essentially consists of a rotor, vanes, ring and a port plate with kidney-shaped inlet and outlet ports. A basic type of vane pump is illustrated in Figure 3.10.

The rotor in a vane pump is connected to the prime mover by means of a shaft. The vanes are placed in radial slots milled into the rotor, which in turn is placed off-center inside a cam ring. When the shaft turns the rotor, the vanes are thrown outward against the cam ring by the centrifugal force and track along the ring, thereby providing a hydraulic seal. The expansion of volume capacity which is indicated by the extension of the vanes as they move through the inlet, allows the atmospheric pressure to push fluid into the pump. The fluid is carried around to the outlet by the vanes whose retraction causes the fluid to be expelled. The pump capacity in vane pumps is determined by vane throw, vane cross-sectional area and the speed of rotation. Vane pumps usually have lower capacity and pressure ratings than gear pumps. But one of the biggest advantages with vane pumps is the high volumetric efficiencies achieved on account of reduced leakage.

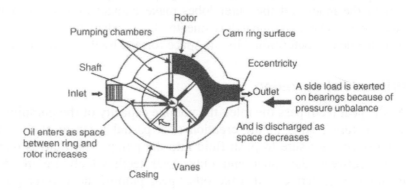

Figure 3.10
Operation of a vane pump

Detailed pump operation

Since the rotor is positioned off-center to the ring during one half of a rotor revolution, the volume between the rotor and the ring increases, thereby permitting the atmospheric pressure acting on the oil tank to force the oil into the pump. This forms the suction process.

During the second half of the rotor revolution, the ring surface will push the vanes back into the rotor slots, and the volume is reduced. This positively ejects the fluid trapped between the ring and the rotor to the outlet port.

Since there are no ports in the ring, a port plate is used to separate the incoming fluid from the outgoing fluid. The port plate fits over the ring, rotor and vanes. The inlet port of the port plate is located at the place where increasing volume is generated and the outlet port at the place where decreasing volume is generated.

All fluids enter and exit the pumping mechanism through the port plate. The off-center distance between the cam and rotor is known as eccentricity. When the eccentricity is zero, there will be no flow.

The following analysis and nomenclature is applicable to a vane pump:
Let
  D_c be the diameter of the cam ring
  D_r be the inside diameter of the rotor
  L be the width of the rotor
  V_d be the displacement volume of the pump
  N be the speed of the pump
  E_{max} be the maximum possible eccentricity
  E be the eccentricity and
  V_{dmax} be the maximum possible volumetric displacement.

The maximum possible eccentricity is given by:

$$E_{max} = \frac{(D_c - D_r)}{2}$$

The maximum possible volumetric displacement is given by

$$V_{dmax} = \frac{\pi}{4}\left(D_c^2 - D_r^2\right)L$$

The actual Volumetric displacement occurs when $E_{max} = E$.
From the above equations

$$V_d = \frac{\pi}{4}\left(D_c + D_r\right) \times E \times L$$

Classification of vane pumps

Vane pumps can further be classified as:

- Balanced vane pumps and
- Variable displacement vane pumps.

Let us discuss the above two categories in detail.

3.5.1   Balanced vane pump

In the unbalanced vane pump, one half of the pumping mechanism experiences pressure that is less than atmospheric, while the other half is subjected to full system pressure. This difference in pressure between the outlet and inlet ports tends to create a severe load on the vanes which along with a large side load on the rotor shaft can lead to bearing failure. It is with a view towards compensating for this deficiency that the concept of balanced vane pump was thought of. The functions of a balanced vane pump are similar to the unbalanced pump except that in the former, the cam ring is cam-shaped or elliptical along with the presence of two intake and two outlet ports that are connected inside the housing.

In our discussion on unbalanced vane pumps, we find that two very different pressures are involved during the course of the pump operation. During one half of the rotor revolution, there is a negative pressure and during the other half revolution of the rotor, the pump is subjected to full system pressure. This results in the side loading of the shaft, which could be severe when high system pressures are encountered. In order to compensate for this, the ring is changed from circular to elliptical shape. The two high-pressure outlet ports are located 180° apart. Because they are located on the opposite sides of the housing, excessive force or pressure buildup on one side is canceled out by equal but opposite forces on the other side. Since the forces acting on the shaft are balanced, the shaft side load is eliminated. In other words, the two pressure quadrants oppose each other and the forces acting on the shaft are balanced. Flow in a balanced vane pump is created very much in the same manner as an unbalanced vane pump, except that there are two suction and discharge cavities instead of one.

A balanced vane pump consists of a cam ring, rotor, vanes and a port plate with inlet and outlet ports opposing each other as shown Figure 3.11. The advent of balanced type vane pumps has led to a significant increase in bearing lifetime, almost about 10–20 times over that of the unbalanced type, in addition to significantly increasing the pressure and speed ratings of these units. Another important point to be noted is that a majority of the constant volume, positive displacement vane pumps used in industrial applications are generally of the balanced type.

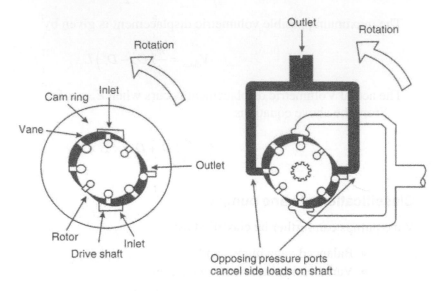

Figure 3.11
Principle of a balanced vane pump

Vane loading

For a vane pump to operate satisfactorily, it is extremely important for a positive seal to exist between the vane tip and the cam ring. The vane pump relies on the centrifugal force to draw the vanes outward against the cam ring to achieve a positive seal. For this reason the minimum operating speed for most vane pumps is 600 rpm.

As the system pressure rises, a tighter seal is required between the vane tip and cam ring in order to minimize internal leakage. For achieving better sealing at high pressures, industrial vane pumps direct system pressure to the underside of the vane. Although this ensures good positive sealing, there is a major disadvantage. By hydraulically loading the vane, the force with which the vane tracks along the cam ring becomes proportional to the discharge pressure. So, at times when the pressure is too high, the force loading the vane is excessive, leading to premature wear of the cam ring and vanes. As a compromise between achieving the best sealing and the least wear, the vanes are only partially loaded.

One way of eliminating this vane loading is by the use of vanes with a chamfer or a beveled edge. Such an arrangement results in the complete underside of the vane area along with a fairly large portion of vane top area, getting exposed to the system pressure. This results in a balance. The pressure which acts on the unbalanced area is in turn, the force loading the vane. The use of bevel-edged vanes has its limitations and is not suitable for application in high-pressure pumps.

3.5.2 Variable displacement vane pump

A positive displacement vane pump delivers the same volume of fluid for each revolution. Industrial pumps generally operate at speeds of 1200 or 1800 rpm indicating the fairly constant nature of the pump flow rate.

The flow rate from the pump in most hydraulic applications needs to be variable, more often than not. One way of achieving this is by varying the speed of the prime mover. However this is not economically practical and therefore, the only other way this can be done is by changing the pump displacement.

The amount of fluid that is displaced by a vane pump is determined by the difference between the maximum and minimum distances by which the vanes are extended and the vane width. While the pump is in operation, we can do nothing, to change the width of the vane. However, the distance by which the vanes are extended can be varied. This is done by making a provision for altering the position of the cam ring relative to the rotor. Such an arrangement enables us to achieve a variable volume from a vane pump. These pumps are known as variable displacement vane pumps. The pumping mechanism of a variable volume vane pump basically consists of a rotor, vanes, cam ring, port plate, thrust bearing for guiding the cam ring and a pressure compensator or hand wheel by which the position of the cam ring can be varied relative to the rotor.

In Figure 3.12, the provision made for changing the position of the cam ring is a screw, which can be manually operated.

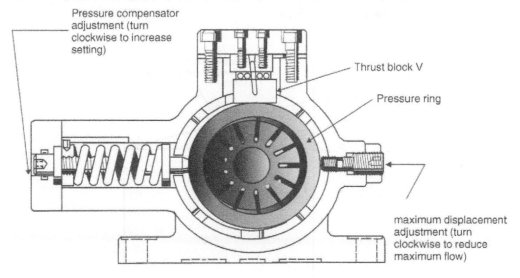

Figure 3.12
Variable displacement pressure compensated vane pump (Courtesy of Brown & Sharpe Mfg. Co, Michigan)

Variable volume vane pumps are unbalanced pumps. Notwithstanding the fact that the rings in these pumps are actually circular and not cam-shaped, they are still referred to as cam-rings.

Working principle

With an inward movement of the screw, the rotor is held off-center with respect to the cam ring (eccentricity). The turning of the rotor induces pumping action. When the screw is drawn outward, the eccentricity is not as much as it was before and therefore the delivery volume is reduced. When the screw is drawn outwards completely, the cam ring centers with the rotor or in other words the eccentricity is zero, resulting in a condition where no increasing or decreasing volume is produced. Hence in spite of the rotor continuing to turn, no pumping action takes place.

3.5.3   Pressure compensation in variable displacement vane pumps

Variable volume vane pumps are generally pressure compensated, which means that when the discharge pressure reaches a certain set value, the pumping action ceases. This is accomplished by using an additional spring called the 'compensator spring' to offset the cam ring.

Initially when the pump discharge pressure is zero, the eccentricity is at a maximum and the spring force will keep the cam ring in the extreme right position. As the discharge pressure increases, it acts on the inner contour of the cam ring and pushes the cam ring towards the left against the spring force, thereby reducing the eccentricity.

When the discharge pressure rises further and becomes high enough to overcome the entire spring force, the compensator spring is compressed until zero eccentricity is achieved. Thus, pumping action ceases and there is no more flow except for minor leakages.

System pressure is therefore limited to the setting of the compensator spring. These pumps therefore ensure their own protection against excessive pressure build-up in the circuit and do not rely on the safety control devices of the hydraulic system.

A flow vs a pressure graph relationship for a variable volume pressure compensated pump is shown in Figure 3.13.

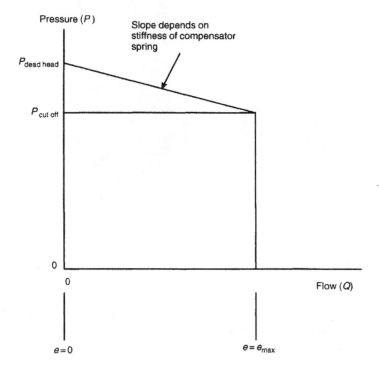

Figure 3.13
Pressure vs flow for pressure compensated vane pump

From the graph, the following can be noticed:

1. Maximum eccentricity gives maximum flow
2. Flow reduces with increase in discharge pressure
3. Zero flow condition results when the eccentricity is zero.

The volumetric efficiency of a good vane pump normally lies between 90 and 95%.

3.6   Piston pump

Piston pumps are high-pressure, high-efficiency pumps. A piston pump works on the principle that a reciprocating piston can draw in fluid when it retracts in a cylinder bore and discharge it when it extends. In other words, these pumps convert the rotary motion of an input shaft to an axial reciprocating motion of the piston. This is usually accomplished with a swash plate that is either fixed or has a variable degree of angle. As the piston barrel assembly rotates, the pistons rotate around the shaft, with the piston slippers contacting the swash plate and sliding along its surface. When the swash plate is vertical, there is no reciprocating motion and hence no displacement occurs. With increase in swash plate angle, there is movement of the pistons in and out of the barrel as it follows the angle of the swash plate surface. During one half of the rotation cycle, the pistons move out of the cylinder barrel, thereby generating an increasing volume. During the other half of the rotation, the pistons move into the cylinder barrel, thereby generating a decreasing volume. This reciprocating motion is responsible for drawing the fluid in and pumping it out.

Piston pumps are basically of two types. They are:

1. Axial piston pumps and
2. Radial piston pumps.

Let us try and understand the design, function and operating principle of each type in detail.

3.6.1   Axial piston pump

Axial piston pumps convert rotary motion of an input shaft to an axial reciprocating motion of the pistons. They in turn are categorized as:

(a) Bent-axis-type piston pumps and
(b) Swash plate-type inline piston pumps.

These two types are discussed separately below.

Bent-axis piston pumps

In these pumps, the reciprocating action of the pistons is obtained by bending the axis of the cylinder block so that it rotates at an angle different than that of the drive shaft. The cylinder block is turned by the drive shaft through a universal link. The centerline of the cylinder block is set at an offset angle, relative to the centerline of the drive shaft. The cylinder block contains a number of pistons along its periphery. These piston rods are connected to the drive shaft flange by ball-and-socket joints. These pistons are forced in and out of their bores as the distance between the drive shaft flange and the cylinder block changes. A universal link connects the block to the drive shaft, to provide alignment and a positive drive. Figure 3.14 shows a bent-axis-type piston pump.

The volumetric displacement of the pump varies with the offset angle 'θ'. There is no flow when the cylinder block centerline is parallel to the drive shaft centerline. 'θ' can vary from 0° to 30°. Fixed displacement units are usually provided with 23° or 30° offset angles.

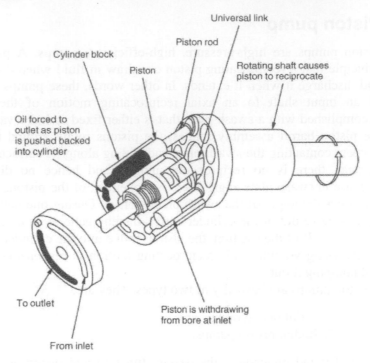

Figure 3.14
Bent-axis piston pump

Variable displacement units are provided with a yoke and an external control mechanism to change the offset angle. Some designs have controls which are capable of moving the yoke over the center position to reverse the direction of flow through the pump.

The following nomenclature and analysis are applicable to an axial piston pump:

Let

θ be the offset angle
S be the piston stroke
D be the piston diameter
Y be the number of pistons and
A be the piston area.

From trigonometry, we have

$$\tan \theta = \frac{S}{D}$$

The total displacement volume equals the number of pistons multiplied by the displacement volume per piston, and is given by

$$V_d = Y \times A \times D \times \tan \theta$$

The flow rate of the pump as calculated from the above equations is given by

$$Q = \frac{(D \times A \times N \times Y \times \tan \theta)}{231}$$

in English units and

$$Q = D \times A \times N \times Y \times \tan \theta \text{ in metric units}$$

Figure 3.15 depicts the typical cross-section of an axial piston pump at different angles.

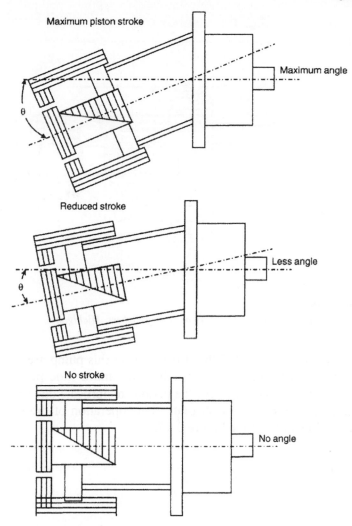

Figure 3.15
Volumetric displacement changes with offset angle

3.6.2   Swash plate inline piston pump

In this type, the axial reciprocating motion of the pistons is obtained by a swash plate that is either fixed or variable in its degree of angle. As the piston barrel assembly rotates, the pistons rotate around the shaft, with the piston shoes in contact with and sliding along the swash plate surface. Since there is no reciprocating motion when the swash plate is in vertical position, no displacement occurs. As there is an increase in the swash plate angle, the pistons move in and out of the barrel as they follow the angle of the swash plate surface. The pistons move out of the cylinder barrel during one half of the cycle of rotation thereby generating an increasing volume, while during the other half of the rotating cycle, the pistons move into the cylinder barrel generating a decreasing volume. This reciprocating motion results in the drawing in and pumping out of the fluid. Pump capacity can easily be controlled by altering the swash plate angle, larger the angle, greater being the pump capacity. The swash plate angle can easily be controlled remotely with the help of a separate hydraulic cylinder. A cross-sectional view of this pump is shown in Figure 3.16.

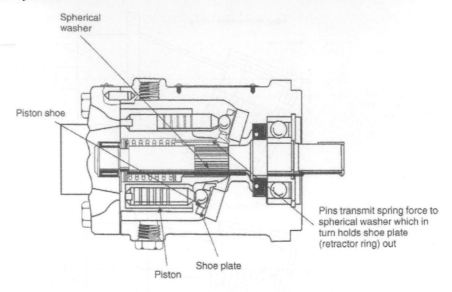

Figure 3.16
Inline design piston pump (Courtesy of Sperry Vickers, Michigan)

The cylinder block and the drive shaft in this pump are located on the same centerline. The pistons are connected through shoes and a shoe plate that bears against the swash plate. As the cylinder rotates, the pistons reciprocate due to the piston shoes following the angled surface of the swash plate. This operation of drawing in and drawing out of the fluid is illustrated in the Figure 3.17.

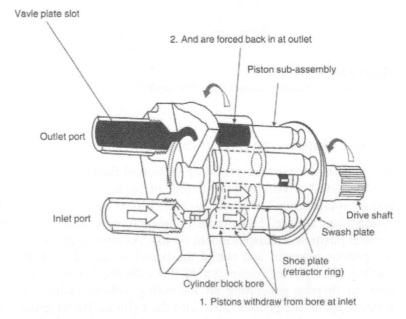

Figure 3.17
Swash plate causes the pistons to reciprocate (Courtesy of Sperry Vickers, Michigan)

The outlet and the inlet ports are located in the valve plate so that the pistons pass the inlet as they are being pulled out and pass the outlet as they are being forced back in.

These types of pumps can also be designed to have a variable displacement capability. In such a design, the swash plate is mounted in a movable yoke. The swash plate angle can be changed by pivoting the yoke on pintles.

The positioning of the yoke can be accomplished by manual operation, servo control or a compressor control and the maximum swash plate angle is usually limited to 17.5° (Figure 3.18).

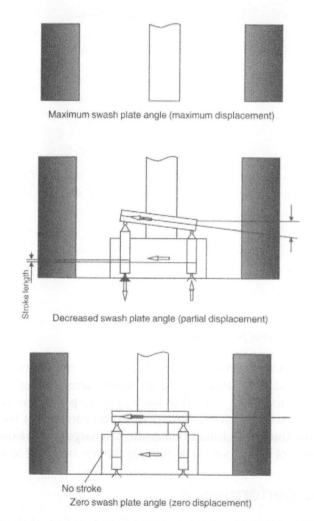

Maximum swash plate angle (maximum displacement)

Decreased swash plate angle (partial displacement)

No stroke
Zero swash plate angle (zero displacement)

Figure 3.18
Swash plate angles

3.6.3　Radial piston pump

These piston pump types have pistons aligned radially on the cylinder block. The operation and construction of a radial piston pump is illustrated in Figure 3.19.

Pump design and operation

The pump consists of a pintle to direct fluid in and out of the cylinder, a cylinder barrel with pistons and a rotor containing a reaction ring. The pistons are placed in radial bores around the rotor. The piston shoes ride on an eccentric ring which causes them to reciprocate as they rotate. The pistons are connected to the inlet port just as they start extending and to the outlet port as they start retracting, by a timed porting arrangement in the pintle. This action is analogous to the brushes and commutator arrangement in a generator. The pistons remain in constant contact with the reaction ring due to the centrifugal force and the back pressure on the pistons.

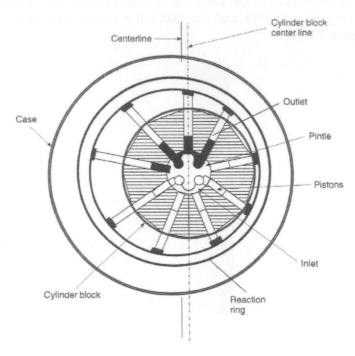

Figure 3.19
Operation of a radial piston pump

For initiating a pumping action, the reaction ring is moved eccentrically with respect to the pintle or shaft axis. As the cylinder barrel rotates, the pistons on one side travel outward. This draws in fluid as each cylinder passes the suction port of the pintle. When the piston passes the maximum point of eccentricity, it is forced inwards by the reaction ring. This forces the fluid to enter the discharge port of the pintle. In certain models, the displacement can be varied by moving the reaction ring to change the piston stroke.

3.7   Pump performance

The performance of a pump is primarily a function of the precision of its manufacturing. Components made to close tolerances should be maintained while the pump is operating under design conditions. Maintenance of close tolerances is achieved by designing systems which incorporate both mechanical integrity and balanced pressures.

Theoretically, an ideal pump is one with zero clearances between all mating parts. Although this is not feasible from the design and manufacturing point of view, the working tolerances should normally be made as small as possible.

Pump manufacturers run tests to determine the performance of their various types of pumps. Overall efficiency of the pumps can be computed by comparing the output of the pump to the power supplied at the input.

Pump efficiencies

The performance data for pumps is determined by a series of tests usually carried out by manufacturers. By comparing the actual hydraulic power output of a pump with the mechanical input power supplied by the prime mover, its overall efficiency can be computed. The overall efficiency can in turn be broken into two distinct components namely, volumetric efficiency and mechanical efficiency.

3.7.1 Volumetric efficiency

This indicates the amount of leakage, which takes place within the pump and involves considerations such as manufacturing tolerances and flexing of the pump casing.

Volumertic efficiency (η_v) is given by

$$\text{Volumetric efficiency} = \frac{\text{Actual flow rate produced by the pump}}{\text{Theoretical flow rate}} \times 100$$

$$= \left(\frac{Q_A}{Q_T}\right) \times 100$$

The volumetric efficiencies for different pump categories are given in the table below.

Type of Pumps	Volumetric Efficiency (%)
Gear pumps	80–90
Vane pumps	82–92
Piston pumps	90–98

3.7.2 Mechanical efficiency

This indicates the amount of energy losses that occur in pumps, not taking into account the leakages. These losses include:

- Friction in bearing and other moving parts and
- Energy losses due to fluid turbulence.

Mechanical efficiency (η_m) is given by the equation,

$$\frac{\text{Pump output power assuming no leakage}}{\text{Actual power delivered to pump}} \times 100$$

$$\eta_m = \frac{\left(\dfrac{P \times Q_T}{1714}\right)}{\left(\dfrac{T_A \times N}{63000}\right)} \times 100 \text{ for power in hp (English units)}$$

Or

$$\eta_m = \frac{(P \times Q_T)}{(T_A \times N)} \times 100 \text{ for power in watts (Metric units)}$$

Where
P is the measured pump discharge pressure in psi or Pa
Q_T is the calculated theoretical pump flow rate in gpm or m³/s
T_A is the actual torque delivered to the pump in in-lbs or N-m
N is the measured pump speed in rpm or rad/s.

Mechanical efficiency can also be computed in terms of torque, i.e.

$$\eta_m = \frac{\text{Theoretical torque required to operate pump}}{\text{Actual torque delivered to pump}} \times 100$$

$$\eta_m = \frac{T_T}{T_A} \times 100$$

One important point to be noted here is that the theoretical torque T_T is the torque required under conditions of no-leakage.

Theoretical torque is given by

$$T_T = \frac{V_D}{2\Pi} \times P$$

Where

T_T is in in.lbs or N-m
V_D is in in.^{3} or m^3 and
P is in psi or Pa.

Actual torque is given by

$$T_A = \frac{\text{Actual hp delivered to pump}}{N \text{ (rpm)}} \times 63\,000$$

Or

$$T_A = \frac{\text{Actual power delivered to pump in W}}{N \text{ (rad/s)}}$$

Where

$$N \text{ (rad/s)} = \frac{2\Pi}{60} \times N \text{ (rpm)}$$

If both volumetric and mechanical efficiencies are known, the overall efficiency can be computed as follows

$$\text{Overall efficiency} = \frac{\text{Actual power delivered by pump}}{\text{Actual power delivered to pump}} \times 100$$

$$\text{Overall efficiency } \eta_o = \frac{(\eta_m \times \eta_v)}{100}$$

Substituting for different values we have,

$$\eta_o = \frac{\dfrac{(P\,Q_A)}{1714}}{\dfrac{(T_A\,N)}{63\,000}} \times 100 \qquad \text{in English units and}$$

$$\eta_o = \frac{(P\,Q_A)}{(T_A\,N)} \times 100 \qquad \text{in metric units and}$$

The actual power delivered to a pump from a prime mover through a rotating shaft is known as Brake power and the actual power delivered by a pump to a fluid is known as hydraulic power.

3.7.3　Pump capacity

Pump manufacturers usually specify pump performance characteristics in the form of graphs, for better visual interpretation. The variation in actual pump capacity depends mainly on the following three factors:

- *Discharge pressure*:　The higher the discharge pressure, greater the internal leakage and hence lower is the actual capacity.
- *Running clearances*:　Large clearances mean greater internal leakage.
- *Oil viscosity*:　The use of low-viscosity oil leads to greater internal leakages.

To keep the design consideration common globally, all gear pumps are designed for their rated capacity at a certain constant pressure. When the pressure at the discharge of a pump increases, the flow rate of the pump reduces. Hence, while designing a pumping system, care should be taken to ensure that the discharge line offers the least resistance to pump discharge.

Effect of running clearance on capacity

Clearances exceeding specified tolerance levels tend to adversely affect the pump performance and efficiency through increased leakage. Let us analyze this further, specifically in relation to gear pumps, in order to examine how any dilution in running clearances ends up adversely affecting pump performance.

All gear pumps are manufactured precisely with specific minimum clearances as per design. However, it is not possible to actually have a zero clearance. Phenomena such as wear, scuffing, abrasive friction and rusting are said to be the major causes for increase in this clearance. It has been found that a 0.04 mm side clearance produces eight times as much internal leakage compared to a 0.02 mm side clearance.

A test was carried out wherein the running clearance on a 1500 rpm pump was intentionally varied, to determine its effect on the pump capacity. During the course of these tests the following findings were established.

A change in the side clearance from 0.025 to 0.045 mm resulted in a 20% reduction in the pump capacity. The pump capacity was found to reduce further with a decrease in pump speed from 1500 to 400 rpm, with the other conditions being held constant.

Effect of oil viscosity on capacity

If the viscosity of the oil is high, the pump capacity is found to increase and vice versa. Since the viscosity is a function of temperature, cold oil results in a higher capacity.

For a similar pump as used earlier and for the operating conditions listed below, the following observations were recorded.

(a) A 20 °C rise in oil temperature led to a decrease in pump capacity by 20%.
(b) Upon further lowering of the pump speed, the pump capacity was found to decrease further.

It is therefore clear that the pump capacity is closely related to the temperature of the oil being used. Hence, it is quite imperative for any testing procedure involving the determination of the pump internal condition or its capacity, to specify the oil temperature, which otherwise would render the whole exercise meaningless.

3.7.4　Pump performance curves

The performance characteristics of pumps are usually specified in the form of graphs. The test data obtained from actual performance tests carried out on pumps are graphically represented for better interpretation. We have seen above the effects of various changes on the capacity of a gear pump. Going further, let us now consider the case of a radial piston pump and study its performance characteristics as represented by the graphs shown in Figure 3.20 and depicting the following relationships.

- Change in discharge flow with change in pump speed
- Change in discharge flow with respect to discharge pressure
- Effect of change in speed on pump efficiency.

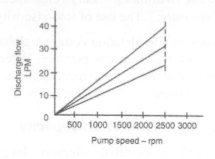

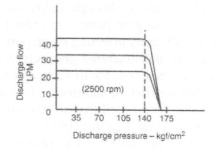

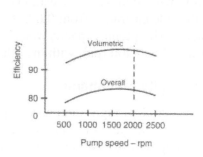

Figure 3.20
Performance curves for radial piston pumps

From the above graphs, the linear relationship between discharge flow and pump speed is established. As regards the relationship between pressure and discharge flow, it is found that the discharge flow is nearly constant over a wide range of pressure. The curves representing volumetric and overall efficiencies are based on a pump whose pressure rating is 2000 psi. Both the efficiencies are found to steadily increase with increase in pump speed before starting to decline at speeds beyond 2000 rpm.

Comparison of pump performance parameters

The table below compares various performance parameters such as pressure, speed, overall efficiency and flow for the various categories of hydraulic pumps.

Type of Pump	Pressure Rating (psi)	Speed Rating (rpm)	Overall Efficiency (%)	Power to Weight Ratio (HP per Pound)	Flow Capacity in gpm
External gear pump	2000–3000	1200–2500	80–90	2	1–150
Internal gear pump	500–2000	1200–2500	70–85	2	1–200
Vane pump	1000–2000	1200–1800	80–95	2	1–80
Radial piston pump	3000–12 000	1200–1800	85–95	3	1–200
Axial piston pump	2000–12 000	1200–3000	90–98	4	1–200

3.7.5   Noise

Noise is another of the important parameters used to determine pump performance. It is measured in units of decibels (dB). Any increase in the noise level normally indicates increased wear and tear and imminent pump failure. Pumps are good generators but poor radiators of noise. However, the noise we hear from the pump in operation is not just the sound coming from the pump, but includes vibration and fluid pulsations as well. In general, fixed displacement pumps are less noisy than variable displacement pumps as they have a rigid construction.

Pump speed has a strong influence on its noise, while the pressure and size have about an equal or lesser effect. The following graph (Figure 3.21) shows these effects.

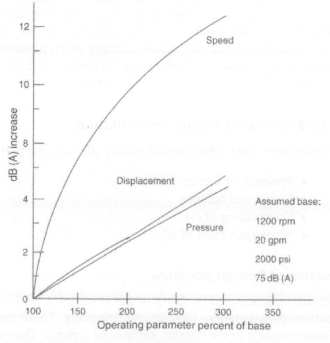

Figure 3.21
Effect of changing pressure, size and speed on noise

Another common cause for noise in hydraulic systems is the presence of entrapped air bubbles in the fluid. These air bubbles, even if they represent less than 1% by volume, change the compressibility of the fluid so much that they can sometimes cause even a fairly silent pump to operate with excessive noise. Entrapped air also leads to the phenomenon of cavitation in pumps about which we shall discuss in detail.

Cavitation in pumps

Pump cavitation is another phenomenon associated with noise and caused by the presence of entrained air bubbles in the hydraulic fluid or vaporization of the hydraulic fluid. It normally occurs when the pump suction lift is excessive and the pump inlet pressure falls below the vapor pressure of the fluid. As a consequence of this, the air bubbles that originate in the low-pressure inlet side of the pump collapse upon reaching the high-pressure discharge side, resulting in increased fluid velocity and giving rise to impact forces which can erode the pump components and shorten its life.

Following are a list of measures that can be undertaken to minimize the effect of cavitation:

    (a) Maintaining short pump inlet lines
    (b) Maintaining suction line velocities below 1.5 m/s (4.9 ft/s)
    (c) Positioning of the pump as close to the reservoir as possible
    (d) Minimizing the number of inlet line fittings
    (e) Using low-pressure drop filters and strainers in the inlet side
    (f) Using the proper oil as per the recommendations of the manufacturer.

3.8   Pump maintenance

As seen from above, pump performance and efficiency are central to efficient transmission of power in a hydraulic system. It therefore becomes necessary to carry out periodic maintenance activities from time to time though it is an established fact that pumps like gear pumps are generally maintenance free within themselves. Let us now discuss in detail, the various factors affecting pump performance, considerations involved in pump selection and also find out how periodic and timely maintenance action can help overcome problems that adversely affect pump performance.

3.8.1   Factors affecting pump performance

Following are some of the external factors affecting the performance of pumps:

- Presence of foreign particles
- Foams and bubbles
- Overheating of oil
- Wrong selection of oil.

Presence of foreign particles

Fluids are contaminated by gritty or metal particles, which come in during the course of extraction, transportation, loading and unloading. These particles get carried in the fluid, and cause damage to the internal surfaces of a pump. They can also accelerate the process of wear and tear of running parts. For example, in gear pumps, the most susceptible components are the side plates and bushings, followed by gear teeth and bearings.

All pumping systems necessarily have a good means of filtration. The filter element provided to ensure clean oil should be periodically replaced. The oil should also be replaced periodically at specified intervals, as it tends to lose its viscosity with prolonged use. The dust in oil promotes oxidation and the oxidized oil in turn adversely affects the hydraulic system.

Foam and bubbles

A noisy pump is often the result of use of foamy oil. Foam and bubbles generate a loud noise when they mesh at the gear teeth. An air bubble between the meshing gear teeth is like a hard rock and can destroy the molecular surface structure of the metal and produce a cavity, often referred to as pitting.

The oil in service might generate foam within itself, particularly when the pressure is low, as in the case of a beer bottle whose cap is prized off the bottle. It entrains external air, with which it may come into contact during re-circulation through the system. Foamy oil is a poor lubricant which reduces pump capacity and promotes oxidation or rusting of metals within the system.

Usually foaming of oil is due to one or more of the following reasons:

- Use of the incorrect type of oil
- A high or low oil level
- Air entering the suction side of pump, which could be due to loose pump bolts or damaged gaskets
- Water mixing with oil.

Overheating of oil

Hot oil is a poor lubricant and increases the internal leakage, thereby reducing pump capacity. It also combines readily with oxygen to accelerate its oxidizing tendency. The temperature limit as regards the phenomenon of overheating is around 90 °C. Sufficient care should be taken to see that the pump is not run at temperatures exceeding 90 °C. High oil temperatures also reduce the life of seals.

Wrong selection of oil

It is important to normally select the oil in accordance with the ambient temperature. This is done in order to enhance the life of the pump components. Furthermore, careful attention is required at the time of adding large quantity of make-up oil. If the condition of the oil in the system is bad, the newly added oil will also get wasted. If the oil in the tank contains soluble sludge, the addition of a large quantity of make-up oil may result in rapid precipitation of this sludge. The normal unwritten rule is that not more than 10% of make-up oil should be added to old oil.

3.8.2 Comparison of various pump performance factors

In general, although gear pumps are the least expensive, they also provide the lowest level of performance. These pumps are simple in design and compact in size. The volumetric efficiency of gear pumps is greatly affected by the leakages that in turn are a result of their constant wear and tear.

Vane pump efficiencies and costs fall between gear and piston pumps. They also last for a longer period.

Although piston pumps are the most expensive of the lot, they provide the highest level of performance. They cannot only be driven at speeds up to 5000 rpm but also operated at very high pressures. Additionally, they are also long lasting, with life expectancy levels of at least seven years. However one major disadvantage with piston pumps is that they cannot be normally repaired in the field on account of their complex design.

The chart below shows a comparison of various performance factors for hydraulic pumps.

Pump Types	Pressure Rating (bar)	Speed (rpm)	Overall Efficiency (%)	HP per lb Ratio	Flow (lpm)
External gear pump	130–200	1200–2500	80–90	2	5–550
Internal gear pump	35–135	1200–2500	70–85	2	5–750
Vane pump	70–135	1200–1800	80–95	2	5–300
Axial pump	135–800	1200–3000	90–98	4	5–750
Radial piston pump	200–800	1200–1800	85–95	3	5–750

3.9   Pump selection

Pumps are selected for a particular application in a hydraulic system based on a number of factors some of which are:

- Flow rate requirement
- Operating speed
- Pressure rating
- Performance
- Reliability
- Maintenance
- Cost and
- Noise.

The selection of a pump typically entails the following sequence of operations:

- Selection of the appropriate actuator (cylinder or motor) based on the load encountered.
- Determining the flow-rate requirements: This involves a calculation to determine the flow rate required to drive the actuator through a specified distance, within a given time limit.
- Determination of the pump speed and selection of the prime mover: This together with the flow rate calculations helps determine the pump size. (volumetric displacement).
- Selection of the pump-type based on the application.
- Selection of system pressure requirements: This also involves determination of the total power to be delivered by the pump.

- Selecting the reservoir capacity along with the associated piping and other related components.
- Computation of the overall system costs.

Normally the above sequence is repeated several times over, for different sizes and types of components. Once this is done, the best overall system is selected for a given application. This process is known as optimization.

3.10 Maintenance practice

Let us now discuss some of the general principles related to maintenance and relevant not only to pumps but also to various other hydraulic/mechanical components.

3.10.1 Cleaning of parts

The usual practice is to thoroughly clean all metal parts with mineral spirit or resort to steam cleaning. It is important to bear the following in mind, during the course of the cleaning process:

- Do not use a caustic soda solution for steam cleaning
- All parts (with the exception of bearings) to be dried with compressed air
- Steam-cleaned parts to be coated with oil (preferably with the same type of oil that is used for the system)
- Ensuring that none of the oil passages are blocked. The passages are to be thoroughly cleaned by working a piece of soft wire back and forth, flushed with mineral spirit and later dried with compressed air.

3.10.2 Inspection of gears

- Inspect the gear for scuffed, nicked, burred or broken teeth. If the defect cannot be removed by a soft honing stone, replacement of the gear needs to be carried out. Additionally, inspect the gear teeth for any damage to the original tooth shape that might have been caused due to wear. If this condition is discovered, get the gear replaced.
- Inspect the thrust face of gears for scores, scratches and burrs. If the defect cannot be removed with a soft honing stone, replace the gear.
- If pitting on the gear tooth is found and if the affected area is more than 1/3rd of the whole area, replace the gear.

3.10.3 Cleaning, installation of seals

Cleaning of seals

The oil seals must be softly wiped with a cloth made wet with petroleum-based solvents, in order to ensure that the tip is not scratched. Care is to be taken to see that solvents such as trichloroethylene, benzol, acetone and all kinds of aromatics which are harmful to seals especially the ones made of polyacrylate rubber are avoided. Oil seals made of quality material like ester rubber may be cleaned by immersing them in light oil. Care must be taken to wipe off the excess oil on the oil seal.

Oil seals must be stored in a location insulated from direct sunlight and moisture. The storage period of oil seals must as a rule be for a maximum of 2 years only.

Oil seal installation

While fitting an oil seal, it is to be ensured that the spring-loaded main lip side faces the oil pressure side (i.e. towards the fluid to be sealed). Also ensure that the surfaces that contact the lip of the seal are free from any roughness, scoring, pitting or wear as this may result in fluid leakage or damage to the seal. When fitting the seal, ensure that it is pressed horizontally. In case it is inclined, the seal has a tendency to get deformed, resulting in oil leakage. Also during installation, no hammer blows should be given on the seals directly. The following tips are quite important in order to ensure longer seal life.

1. *Coating the seal lip with lubricating oil or grease*:   This prevents dry friction during the initial period of the seal's movement.
2. *Coating the circumference of the seal with adhesive or sealant*:   While doing this, it is to be ensured that the adhesive/sealant does not make contact with the lip.

The circumference of some seals is precoated with a dry sealant. The sealant is usually colored for easy identification. Such seals do not require any additional sealant during installation.

3.10.4   Inspecting castings and machined surfaces

1. Inspect the bores for scratches, grooves, burrs and dirt. Ensure removal of scratches and burrs with crocus cloth. Also ensure replacement of the part that has scratches or grooves.
2. Inspect mounting surfaces for nicks, scratches and burrs. Get the same removed by using a soft stone or crocus cloth. In the event of the scratches or burrs being difficult to remove, get the part replaced.
3. Inspect castings or housings that are cracked. All machined surfaces are to be examined on a periodic basis to determine damage if any, that could lead to possible fluid leakage.

3.10.5   Gaskets, seal rings and O-rings

1. As a general rule, all gaskets, seal rings and o-rings in a component that has been disassembled are to be replaced.
2. Both faces of the gasket are to be examined for sticking foreign matter or scratches and the same removed.
3. Both surfaces of the gasket are to be coated with a liquid gasket. Each brand of liquid gasket has its own drying time and therefore all instructions are to be carefully adhered to.
4. To remove a seal ring, install a thin blade into the seal ring groove, and work one edge of the seal ring out in such a manner that it can be grasped with the fingers and removed.
5. Lightly coat the seal ring contact surface with a lubricant during the installation. If lip type seal rings are used, ensure that the main lip faces the direction of the oil pressure.

O-rings are available in various types and sizes depending on their position, temperature and pressure. They are to be chosen accordingly. O-rings are also to be replaced carefully, based on application conditions.

Another important point to be noted is that discarded teflon seal rings when burned, have a tendency to emit toxic poisonous gases and therefore this is best avoided.

4

Hydraulic motors

4.1 Objectives

After reading this chapter, the student will be able to:

- Differentiate between hydraulic pumps and motors
- Understand and describe the design and construction of various motors used in hydraulics
- Explain the operation of the various hydraulic motors of the likes of gear, vane and piston motors and also evaluate their performance by determining their mechanical, volumetric and overall efficiencies
- Select and size motors for hydraulic applications
- Understand the performance parameters of hydraulics motors.

4.2 Basic principles

Hydraulic motors are classified as rotary actuators. But strictly speaking, the term 'Rotary actuator' is reserved for a particular type of unit whose rotation is limited to less than 360°. Hydraulic motors are used to transmit fluid power through linear or rotary motion. They resemble pumps very closely in construction. However, as already understood, pumps perform the function of adding energy to a hydraulic system for transmission to some remote point, while motors do precisely the opposite. They extract energy from a fluid and convert it to a mechanical output to perform useful work. To put it more simply, instead of pushing on the fluid as the pump does, the fluid pushes on the internal surface area of the motor, developing torque. Since both the inlet and outlet ports in a motor may be pressurized, most hydraulic motors are externally drained.

Hydraulic motors can be of limited rotation or continuous rotation type. Limited rotation motors are called oscillation fluid motors as they produce a reciprocating motion. Continuous rotation motors (hydraulic motors) as mentioned earlier are in reality pumps, which have been redesigned to withstand the different forces that are involved in motor applications.

Hydrostatic transmissions are hydraulic systems specifically designed to have a pump drive a hydraulic motor. Thus, a hydraulic transmission system simply transforms mechanical power into fluid power and then reconverts the fluid power into shaft power.

In this chapter, we shall examine the various types of hydraulic motors from the design and construction point of view and also evaluate their performance.

Hydraulic motors can be classified into two types:

1. Limited rotation hydraulic motors
2. Continuous rotation hydraulic motors.

Let us now examine each of these types in detail.

4.3   Limited rotation hydraulic motor

A limited rotation hydraulic motor provides a rotary output motion over a finite angle. This device produces a high instantaneous torque in either direction and requires only a small amount of space and simple mountings.

Rotary motors consist of a chamber or chambers containing the working fluid and a movable surface against which the fluid acts. The movable surface is connected to an output shaft to produce the output motion.

Figure 4.1 shows a direct acting vane-type actuator. In this type, fluid under pressure is directed to one side of the moving vane, causing it to rotate. This type of motor provides about 280° rotation.

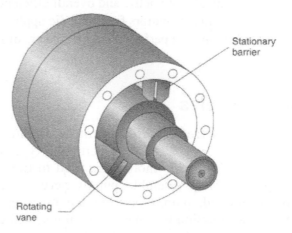

Figure 4.1
Limited rotation hydraulic actuator

Rotary actuators are available with working pressures up to 350 kg/cm² (4978 psi). They are typically foot mounted, flanged or end mounted. Most designs provide cushioning devices. In a double vane design similar to the one depicted in the figure above, the maximum angle of rotation is reduced to about 100°. However in this case, the torque-carrying capacity is twice that obtained by a single vane design.

4.3.1 Torque capacity analysis of limited rotation hydraulic actuators with single rotating vane

Nomenclature

Let

R_R be the outer radius of the rotor in meters
R_V be the outer radius of the vane in meters
L be the width of the vane in meters
P be the hydraulic pressure in psi or Pascal
F be the hydraulic force acting on the vane in Newton
A be the surface area of the vane in contact with oil in meter^2 and
T be the torque capacity in Newton meter.

The force on the vane equals pressure times the vane surface area which is given by:

$$F = P \times A$$
$$= P(R_V - R_R)L$$

The torque equals the vane force times the mean radius of the vane. This is given by:

$$T = \dot{P}(R_V - R_R)L\left\{\frac{(R_V + R_R)}{2}\right\}$$

On rearranging, the equation can be written as:

$$T = \left\{\frac{PL}{2}\right\}(R_V{}^2 - R_R{}^2)$$

Volumetric displacement (V_d) is given by

$$V_D = \pi \ (R_V^2 - R_R^2)\ L$$

Combining the above two equations, we get

$$T = \frac{(PV_D)}{6.28}$$

It is observed from the above equation that torque capacity can be increased either by increasing the pressure or by increasing the volumetric displacement or both.

The various applications of rotary type limited rotation motors are (Figure 4.2):

- Conveyor sorting
- Valve turning
- Air bending operation
- Flip over between workstations
- Positioning for welding
- Lifting, rotating and dumping.

Figure 4.2
Application of limited rotation hydraulic motor

4.4　Continuous rotation hydraulic motors

Continuous rotation hydraulic motors are actuators, which can rotate continuously. Instead of acting on (or pushing) the fluid as pumps do, motors are acted upon by fluids. In this way, hydraulic motors develop torque and produce continuous rotary motion. Since the casing of a hydraulic motor is pressurized from an outside source, most hydraulic motors have casing drains to protect their shaft seals.

These are further classified as:

- Gear motors
- Vane motors and
- Piston motors.

Let us discuss each of the motor types individually.

4.4.1　Gear motors

Gear motors are simple in construction. A gear motor develops torque due to the hydraulic pressure acting on the surfaces of the gear teeth. A cross-section of a typical gear motor is illustrated in Figure 4.3.

By changing the direction of the flow of fluid through the motor, the direction of rotation of the motor can be reversed. As in the case of a gear pump, the volumetric displacement of the motor is fixed. The gear motor is not balanced with respect to pressure loads. The high pressure at the inlet, coupled with the low pressure at the outlet, produces a large side load on the shaft and bearings, thereby limiting the bearing life of the motor.

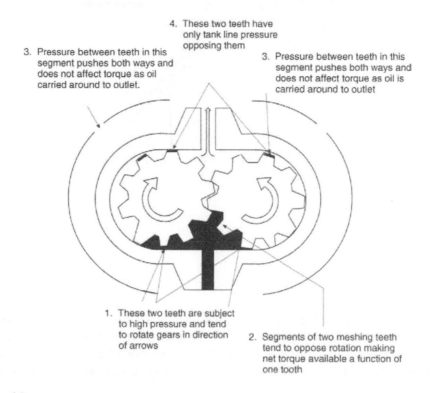

4. These two teeth have only tank line pressure opposing them

3. Pressure between teeth in this segment pushes both ways and does not affect torque as oil carried around to outlet.

3. Pressure between teeth in this segment pushes both ways and does not affect torque as oil is carried around to outlet

1. These two teeth are subject to high pressure and tend to rotate gears in direction of arrows

2. Segments of two meshing teeth tend to oppose rotation making net torque available a function of one tooth

Figure 4.3
Torque development by a gear motor

Gear motors are normally limited to operating pressures of around 140 kg/cm^2 (roughly 2000 psi) and operating speeds of 2400 rpm. They are available with a maximum flow capacity of 550 lpm (liters per minute).

Hydraulic motors can also be of the internal gear type. The internal gear type motors can operate at higher speeds and pressures. They also have greater displacements than the external motors. Screw-type motors also form part of gear motors. As in the case of pumps, screw-type hydraulic motors use three meshing screws. The rolling screw set results in an extremely quiet operation. Screw type motors can operate at pressures up to 210 kg/cm^2 (2990 psi approximately) and can have displacement volumes up to 0.227 liter.

The main advantages associated with gear motors are its simple design and cost-effectiveness. They also possess good tolerance to dirt. The main disadvantages with gear motors are their lower efficiency levels and comparatively higher leakages.

4.4.2 Vane motors

The internal construction of the vane motors is similar to that of a vane pump; however the principle of operation differs.

Vane motors develop torque by virtue of the hydraulic pressure acting on the exposed surfaces of the vanes, which slide in and out of the rotor connected to the drive shaft.

As the rotor revolves, the vanes follow the surface of the cam ring because springs are used to force the vanes radially outward. Figure 4.4 illustrates the basic operation of a vane motor.

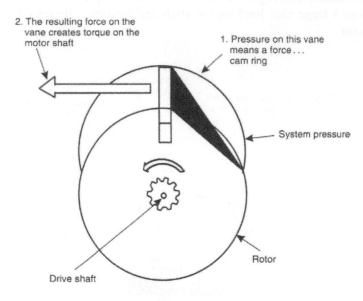

Figure 4.4
View of basic operation

No centrifugal force exists until the rotor starts to revolve. Therefore, the vanes must have some means other than the centrifugal force to hold them against the cam ring. Some designs use springs, while other types use pressure-loaded vanes. The sliding action of the vanes forms sealed chambers, which carry fluid from the inlet to the outlet.

Balanced and unbalanced-type vane motors

Figure 4.5(a) shows an unbalanced-type vane motor comprising of a circular chamber having an eccentric rotor and carrying several spring or pressure-loaded vanes. A higher force is exerted by the fluid on the upper vanes since fluid flowing through the inlet port finds more area of the vanes exposed in the upper half of the motor. This results in the counter-clockwise rotation of the rotor. The motor displacement here is a function of its eccentricity.

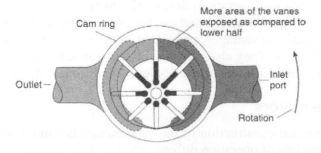

Figure 4.5(a)
Unbalanced vane motor

In the unbalanced type, as all the inlet pressure acts on one side of the rotor, the radial load on the shaft bearing is quite large. This problem is alleviated to a large extent by using a balanced vane motor, a schematic of which is illustrated in Figure 4.5(b).

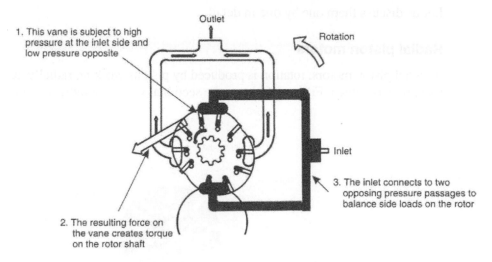

Figure 4.5(b)
View of balanced design operation

In the balanced type, a double-lobed ring with diametrically opposed ports is provided. Here the side force that is exerted on one side of the bearing is canceled by an equal but opposite force from the pressure port located diametrically opposite. The like ports are usually connected internally so that only one inlet and outlet ports are brought outside.

Vane motors are universally of the balanced design type. Since vane motors are hydraulically balanced, they are fixed displacement units. These motors can operate at pressures of up to 175 kg/cm^2 (2490 psi) and speeds up to 4000 rpm. The maximum flow usually delivered by these motors is in the range of 950 lpm. The vane-type motors have more internal leakage as compared with the piston type and are therefore not recommended for use in servo control systems.

4.4.3 Piston motors

Piston motors are also similar in construction to that of piston pumps. Piston motors can be either fixed or variable displacement units. They generate torque through pressure acting at the ends of pistons, reciprocating inside a cylinder block. To put it rather simply, piston-type hydraulic motors use single-acting pistons that extend by virtue of fluid pressure acting on them and discharge the fluid as they retract. The piston motion is translated into circular shaft motion by different means such as an eccentric ring, bent axis or with the help of a swash plate.

The piston motor design usually involves incorporation of an odd number of pistons. This arrangement results in the same number of pistons receiving the fluid as the ones discharging the fluid, although one cylinder may get blocked by the valve crossover. On the contrary, with the number of pistons being even and one getting blocked, there would be one more piston either receiving or discharging the fluid leading to speed and torque pulsations.

Piston motors are the most efficient of all motors. They are capable of operating at very high speeds of 12 000 rpm and also pressures up to 350 kg/cm^2 (4980 psi approx.). Large piston motors are capable of delivering flows up to 1500 lpm.

Piston motors are further categorized as:

1. Radial piston motors and
2. Axial piston motors.

Let us discuss them one by one in detail.

Radial piston motors

In radial piston motors, rotation is produced by pistons working radially against an eccentric track ring as shown. Figure 4.6 shows the sectional view of a radial piston motor.

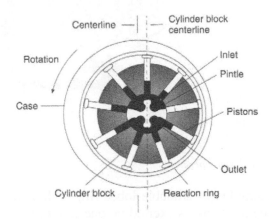

Figure 4.6
Radial piston motor

By directing the fluid to half of the cylinder bores, the pintle functions like a stationary hydraulic valve and forces the pistons outward radially. The other side of the pintle allows the fluid to be discharged from the retracting pistons. Pistons under pressure move by revolving towards the point where the track ring is farthest from the cylinder block axis.

The cylinder block and the shaft are carried by the pistons as they rotate, all the while ensuring that the pintle is lined up with the proper bores. The pistons that are not exposed to the pressure of the inlet fluid, move towards the cylinder block axis and in the process expel return oil.

The displacement in a radial hydraulic motor is a function of the total piston area as well as the eccentricity of the track ring. Radial motor applications are generally limited to higher horsepower units.

Axial piston motors

In an axial piston motor, the rotor rotates on the same axis as the pistons. There are basically two types of axial piston motor design. They are:

1. In-line piston motor (swash plate type) and
2. Bent-axis type.

In-line piston motor

Figure 4.7 illustrates an in-line design in which the motor drive shaft and cylinder block are centered on the same axis. Here hydraulic pressure acting at the ends of the pistons generates a reaction against an angled stationary swash plate. This causes the cylinder block to rotate with a torque that is proportional to the area of the pistons. The torque is also a function of the swash plate angle. The in-line piston is designed either as a fixed or variable displacement unit. The swash plate angle generally determines the volumetric displacement.

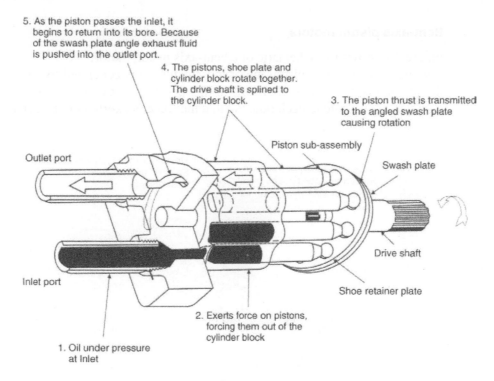

5. As the piston passes the inlet, it begins to return into its bore. Because of the swash plate angle exhaust fluid is pushed into the outlet port.

4. The pistons, shoe plate and cylinder block rotate together. The drive shaft is splined to the cylinder block.

3. The piston thrust is transmitted to the angled swash plate causing rotation

Piston sub-assembly

Swash plate

Outlet port

Drive shaft

Inlet port

Shoe retainer plate

2. Exerts force on pistons, forcing them out of the cylinder block

1. Oil under pressure at Inlet

Figure 4.7
In-line piston motor operation

In variable displacement units, the swash plate is mounted in a swinging yoke. The angle of the swash plate can be altered by various means such as a lever hand wheel or a servo control. If the swash plate angle is increased, the torque capacity is increased, but the drive shaft speed is decreased. Mechanical stops are usually incorporated so that the torque and speed capacities stay within prescribed limits. The operation of a variable displacement unit is illustrated in the Figures 4.8(a) and (b).

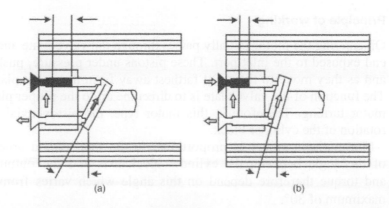

(a)   (b)

Figure 4.8
Motor displacement variation with the swash plate angle

The cylinder block in an in-line piston motor rotate with a torque that is proportional to the area of the pistons. The torque is also a function of the swash plate angle. The in-line piston is designed either as a fixed or variable displacement unit. The swash plate angle generally determines the volumetric displacement.

Bent-axis piston motors

Figure 4.9 shows the schematic of a bent-axis piston motor.

In the bent-axis type piston motor too, the torque is generated by the pressure acting on the reciprocating pistons. However in this design, the cylinder block and drive shaft are mounted at an angle to each other so that the force is exerted on the drive shaft flange.

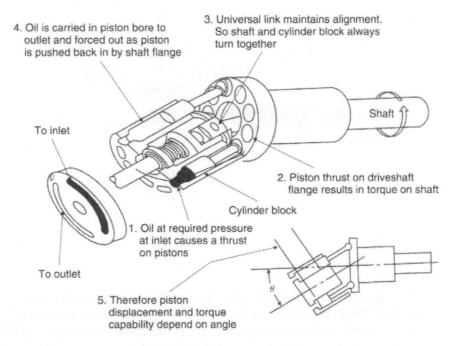

4. Oil is carried in piston bore to outlet and forced out as piston is pushed back in by shaft flange

3. Universal link maintains alignment. So shaft and cylinder block always turn together

To inlet

Shaft

2. Piston thrust on driveshaft flange results in torque on shaft

Cylinder block

1. Oil at required pressure at inlet causes a thrust on pistons

To outlet

5. Therefore piston displacement and torque capability depend on angle

θ

Figure 4.9
Bent-axis piston motor (Courtesy of Sperry Vickers, Michigan)

Principle of working

Oil entering the motor initially passes through the valve plate and then enters at the piston end exposed to the inlet port. These pistons under pressure, push against the shaft flange and as they move to the point farthest away from the valve plate, rotate the drive shaft. The function of the valve plate is to direct the oil to the proper pistons in order to keep the motor turning. Therefore in this motor type, the piston axial travel is responsible for rotation of the cylinder block.

Motor displacement is proportional to the total piston area and varies as the sine of the angle between the cylinder block axis and the output shaft axis. The speed and torque therefore depend on this angle which varies from a minimum of 7° to a maximum of 30°.

With a fixed inlet flow, the smaller this angle, the higher the speed and with a fixed maximum pressure, the smaller the angle, smaller is the maximum torque.

The bent-axis piston motor is a highly efficient motor and is found to have very low leakage levels compared with other motors. It has been used extensively in aircraft servo systems, although the latest design in-line motors have largely replaced them in recent times.

From a study of both the in-line and bent-axis motors, it is found that the characteristics of the in-line motor are very similar to those of the bent-axis type with the swash plate angle replacing the axis angle. The displacement in either motor can be varied by changing this angle. This notwithstanding, one major difference between the two is the reduced torsional efficiency of the in-line unit caused by the frictional drag of its shoe plate against the swash plate. Although this leads to a decrease in output horsepower along with a temperature rise, the added friction increases damping to the servo loop. The loss in torsional gain can be overcome by the inherently high-pressure gain of a normal servo valve.

The frictional losses in a hydraulic motor result in the motor delivering less torque than it should theoretically. The theoretical torque therefore is that value of torque that is delivered by a frictionless motor and can be determined by the equation for limited rotation hydraulic actuators discussed earlier, which is

$$T = \frac{PV_D}{6.28}$$

Where

P is the pressure in Pascal and

V_D is the volumetric displacement measured in m^3/rev.

Thus theoretical torque is not only proportional to the pressure but also to the volumetric displacement. The theoretical power output, i.e. the power developed by a frictionless motor can be mathematically expressed as

$$Hp = \frac{T \times N}{63\,000}$$

$$= \frac{(P \times V_D \times N)}{(6.28 \times 63\,000)}$$

As is the case with pumps, the relationship between speed, volumetric displacement and flow is given by:

$$Q = \frac{(VD \times N)}{231}$$

4.5   Hydraulic motor performance

The performance of a hydraulic motor depends on a lot of factors such as:

- Manufacturing precision
- Maintenance of close tolerances
- Internal leakage
- Friction between mating parts and
- Internal fluid turbulence.

While internal leakage between the motor inlet and outlet leads to decreased volumetric efficiency, loss in mechanical efficiency occurs on account of friction between the mating parts and also due to fluid turbulence.

The table below shows the typical overall efficiencies for gear, vane and piston motors.

Type of Motor	Efficiency (%)
Gear motors	70–75
Vane motors	75–85
Piston motors	85–95

Hydraulic motor performance is evaluated on the basis of the same basic parameters of volumetric efficiency, mechanical efficiency and overall efficiency, similar to that of hydraulic pumps.

4.5.1   Volumetric efficiency

The volumetric efficiency (η_v) of a hydraulic motor is given by

$$\eta_v = \left(\frac{Q_T}{Q_A}\right) \times 100$$

Where
Q_T is equal to the theoretical flow rate the motor should consume and
Q_A is the actual flow rate consumed by the motor.

From the above equation, the volumetric efficiency of a hydraulic motor is found to be the exact inverse of that of a hydraulic pump. The reason for this being that a motor consumes more flow than it should theoretically, on account of leakage, whereas a pump does not produce as much flow as it should theoretically.

The volumetric efficiency can be determined, if the values of Q_T and Q_A are known. While Q_T, the theoretical flow rate can be determined with the help of the equation derived earlier, Q_A, the actual flow rate is measured.

4.5.2   Mechanical efficiency

The mechanical efficiency (η_m) of a hydraulic motor is given by

$$\frac{\text{Actual torque delivered by the motor}}{\text{Torque which the motor should theoretically deliver)}} \times 100$$

$$\eta_m = \left(\frac{T_A}{T_T}\right) \times 100$$

Where
T_A is the actual torque delivered by the motor and
T_T is the theoretical motor torque.

The mechanical efficiency can be determined, if the values of T_A and T_T are known.
T_T is given by

$$\frac{V_D \times P}{6.28}$$

Where

T_T is in in-lbs or N-m

V_D is in in^3 or m^3/rev and

P is in psi or Pa and

T_A is given by:

$$T_A = \frac{\text{Actual motor hp} \times 63\,000}{N} \quad \text{in English units}$$

Or

$$T_A = \frac{\text{Actual motor wattage}}{N} \quad \text{in metric units}$$

From the above equations, the mechanical efficiency of a hydraulic motor is again found to be the exact inverse of that of a hydraulic pump. This is because a pump requires greater torque than it should theoretically, on account of friction, whereas lesser torque is produced by the motor than it should theoretically.

Overall efficiency

The overall efficiency (η_o) for motors as in the case of pumps is given by the product of the volumetric and mechanical efficiencies.

$$\text{Overall efficiency} = \frac{\text{Actual power delivered by the motor}}{\text{Actual power delivered to the motor}}$$

$$\eta_o = \frac{(\eta_v \times \eta_m)}{100}$$

$$\eta_o = \frac{T_A \times N}{P \times Q_A}$$

Where

T_A is in Newton meter

N is in rad/s

P is in Pascal and

Q_A is in m^3/s.

Or

$$\eta_o = \left(\frac{T_A \times N}{63\,000}\right)$$

$$= \left(\frac{P \times Q_A}{1714}\right)$$

Where

T_A is in in-lbs

N is in rpm

P is in psi

Q_A is in gpm.

One important point to be noted is that while the actual power delivered to a motor by the fluid is known as h*ydraulic power*, the actual power delivered by the motor to a load through a rotating shaft is known as *brake power*.

4.6 Electro-hydraulic stepping motors

An electro-hydraulic stepper motor (EHSM) is a device, which uses a small electrical stepper motor to control the huge power available from a hydraulic motor (Figure 4.10).

It consists of three components:

1. Electrical stepper motor
2. Hydraulic servo valve
3. Hydraulic motor.

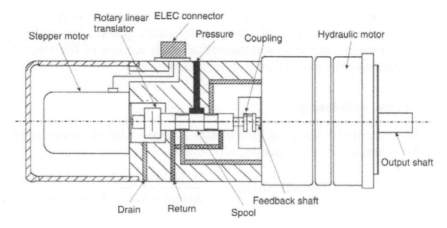

Figure 4.10
The electro-hydraulic stepping motor (Courtesy of Motion Products, Minnesota)

These three independent components when integrated in a particular fashion provide a higher torque output, which is several hundred times greater than that of an electrical stepper motor.

The electric stepper motor undergoes a precise, fixed amount of rotation for each electrical pulse received. This motor is directly coupled to the rotary liner translator of the servo valve. The output torque of the electric motor must be capable of overcoming the flow forces in the servo valve. The flow forces in the servo valve are directly proportional to the rate of flow through the valve. The torque required to operate the rotary linear translator against this axial force is dependent on the flow gain in the servo valve.

The hydraulic motor is the most important component of the EHSM system. The performance characteristics of the hydraulic motor determine the performance of the EHSM. These are typically used for precision control of position and speed. These motors are available with displacements ranging from 0.4 cubic in. (6.5 cm^3) to 7 cubic in. (roughly 115 cm^3). Their horsepower capabilities range between 3.5 hp (2.6 kW) and 35 hp (26 kW). Typical applications include textile drives, paper mills, roll feeds, automatic storage systems, machine tools, conveyor drives, hoists and elevators.

Hydraulic cylinders

5.1 Objectives

After reading this chapter, the student will be able to:

- Explain the construction and design features of hydraulic cylinders
- Describe in detail the operating principles of cylinders
- Explain the construction of various types of cylinders used in hydraulic systems
- Calculate the various cylinder performance parameters such as load-carrying capacity, speed and power
- Select and size cylinders for hydraulic applications
- Troubleshoot common cylinder problems.

5.2 General

A hydraulic system is generally concerned with activities related to moving, gripping or applying force to an object. Devices, which achieve these objectives, are referred to as actuators. Actuators are interface components that convert hydraulic power back to mechanical power. Based on whether an actuator gives rotational motion or linear motion, actuators are basically categorized as:

- Rotary actuators and
- Linear actuators.

Rotary actuators are nothing but motors that we have examined in the previous chapter. Linear actuators, as the name implies, are used to move objects or apply a force in a straight line. These are otherwise known as hydraulic cylinders.

Cylinders are linear actuators whose output force or motion is in a straight line. Their function is to convert hydraulic power into linear mechanical power. Hydraulic cylinders extend and retract to perform a complete cycle of operation. Their work applications as earlier discussed may include pulling, pushing, tilting and pressing. The type of cylinder to be used along with its design is based on a specific application. The simplest of linear actuators is a ram which is shown in Figure 5.1. It has only one fluid chamber and exerts force in one direction only. Rams are widely used in applications where stability is needed on heavy loads. Ram-type cylinders are practical for long strokes and are used on jacks, elevators and automobile hoists.

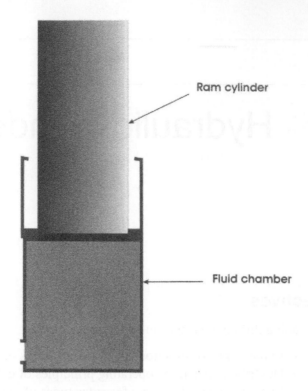

Ram cylinder

Fluid chamber

Figure 5.1
Hydraulic ram

Hydraulic cylinders are further classified as:

- Single acting cylinders and
- Double acting cylinders.

Let us discuss these two cylinder types in detail from the design, construction and application point of view.

5.3   Construction

5.3.1   Single acting cylinder

Single acting cylinders are pressurized at one end only while the opposite end is vented to the atmosphere or tank. They are usually designed in such a way that a device such as an internal spring retracts them. Figure 5.2 is an illustration of the simplest form of a single acting hydraulic cylinder along with its symbolic representation.

A single acting hydraulic cylinder consists of a piston inside a cylindrical housing called a barrel. Attached to one end of the piston is a rod, which extends outside one end of the cylinder (rod end). At the other end (blank end) is a port for the entrance and exit of oil.

A single acting cylinder is pressurized at one end only i.e. it can exert a force in only the extending direction as the fluid from the pump enters the blank end of the cylinder. The opposite end is vented to the tank or atmosphere or in other words, these cylinders do not retract hydraulically. Retraction is accomplished by using gravity or by the inclusion of a compression spring at the rod end.

The force applied by the piston depends on both the area and the applied pressure. This force applied for a given cylinder can be calculated by the equation:

$$F = P \times \pi \times r^2$$

Where
> F is the force applied in kgf or N
> P is the pressure in psi or Pascal and
> r is the radius of the cylinder in foot or meter.

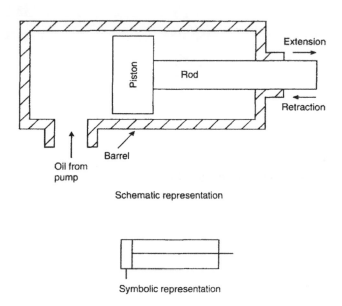

Schematic representation

Symbolic representation

Figure 5.2
Single acting hydraulic cylinder

5.3.2 Double acting cylinder

Double acting cylinders are the most commonly used cylinders in hydraulic applications. Here pressure can be applied to either port giving power in both directions. Due to the fact that these cylinders have unequal exposed areas during extend and retract operations, they are also known as differential cylinders. This difference in effective area is in turn caused by the area of the rod that reduces the piston area during retraction. Since more fluid is required to fill the piston side of the cylinder during extension, the operation is understandably slower. However because of the greater effective area, more force is generated on extension. During the retraction operation, the same amount of pump flow will retract the cylinder faster because of the reduced fluid volume displaced by the rod. However, the reduced effective area during this operation results in the generation of less force.

Figure 5.3 shows the typical construction of a double acting cylinder.

The cylinder consists of five basic parts: two end caps (a base cap and a bearing cap) with port connections, a cylinder barrel, a piston and the rod itself. This basic construction provides for simple manufacture as the end caps and pistons remain the same for different lengths of the same diameter cylinder. The end caps can be secured to the barrel by welding, through tie rods or by threaded connections.

An enlarged view of the bearing cap welded to the barrel has been illustrated in Figure 5.4.

The inner surface of the barrel needs to be very smooth to prevent wear and leakage. Generally, a seamless drawn steel tube machined to an accurate finish is used. In applications where the cylinder is used infrequently or involves possible contact with corrosive materials, stainless steel, aluminum or brass may be used as the cylinder material.

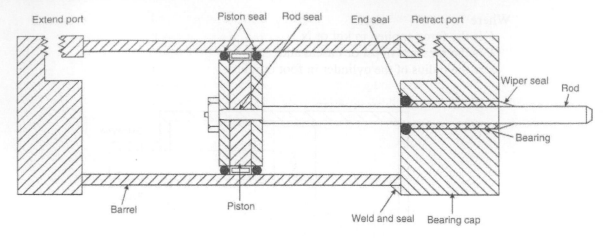

Figure 5.3
Construction of a double acting cylinder

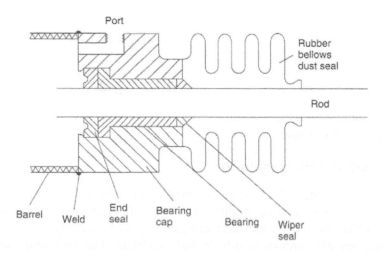

Figure 5.4
Enlarged view of the bearing cap welded to a barrel

Pistons are usually made of cast iron or steel. The piston not only transmits force to the rod, but must also act as a sliding bearing in the barrel and provide a seal between the high- and low-pressure sides. Piston seals are generally used between the piston and barrel. Occasionally when small leakages can be tolerated, piston seals are not used. The usual practice is to deposit a bearing surface such as bronze onto the piston surface which is then honed to a finish, similar to that of the barrel.

The surface of the cylinder rod is exposed to the atmosphere when extended and hence is liable to suffer from the effects of dirt, moisture and corrosion. When the cylinder is retracted, contaminants of the likes of dirt, moisture and corrosive materials may be drawn back into the barrel, causing problems inside the cylinder. Heat-treated chromium alloy steel is generally used for added strength. It also doubles up as a counter against the harmful effects of corrosion.

In order to remove dust particles, these cylinders are fitted with a wiper or scraper seal fitted to the end cap. In dusty atmospheres, external rubber bellows may also be used to prevent the entry of dust. This mounting arrangement of the rubber bellow seal is shown in the Figure 5.3 above.

In order to prevent the leakage of fluid on the high-pressure side along the rod, an internal sealing ring is fitted behind the bearing. In some designs, the wiper seal and the bearing and sealing rings are combined to form a cartridge assembly in order to simplify maintenance. The rod is normally attached to the piston via a threaded end. In order to prevent leakage along the rod, cap seals are provided. These combine the roles of piston and rod seals or act as a static O-ring around the rod as shown in the Figures 5.5(a) and (b).

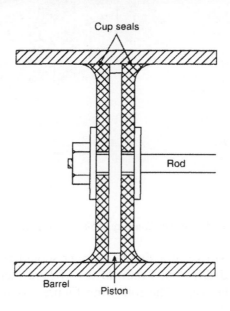

Figure 5.5(a)
Cup seals

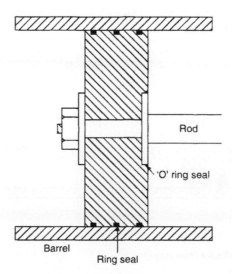

Figure 5.5(b)
Ring seals and O-ring

The nomenclature used for a typical double acting cylinder has been illustrated in Figure 5.5(c) which is the cross-sectional view of a double acting cylinder manufactured by Sheffer Corporation, Ohio.

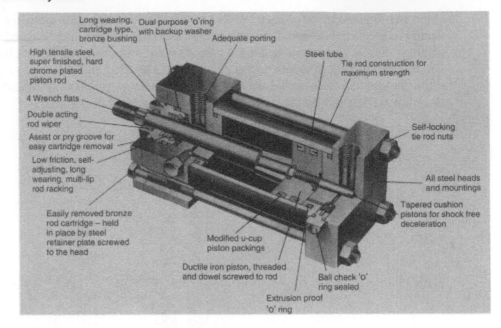

Figure 5.5(c)
Double acting cylinder design (Courtesy of Sheffer Corporation, Ohio)

Ideally, the stroke of a simple cylinder should be less than the barrel length, giving at best an extended/retracted ratio of 2:1. Where there is restriction in space, a telescopic cylinder can be used. The following Figure 5.6 shows a typical double acting cylinder with two pistons.

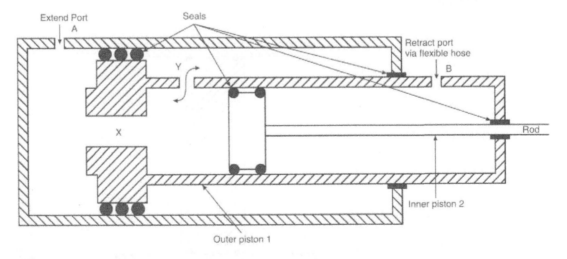

Figure 5.6
Double acting cylinder (two stage)

To extend the rod, fluid pressure is applied to port A. It is applied on both sides of the piston 1 via ports X and Y, but the difference in areas between the sides of piston 1, causes the piston to move to the right.

To retract the rod, fluid is applied to port B. When piston 2 is fully driven to the left, the port Y gets connected to port B, applying pressure to the right side of the piston 1, which then retracts.

5.4   Cylinder cushioning

The end caps are generally made of cast iron or cast aluminum and incorporate threaded entries for ports. End caps have to withstand shock loads at the extreme ends of piston travel. These loads arise not only from fluid pressure, but also from the kinetic energy of the moving parts of the cylinder and load. These shock loads can be reduced with cushion valves built in the end caps.

The cylinder cushioning effect has been illustrated in Figure 5.7.

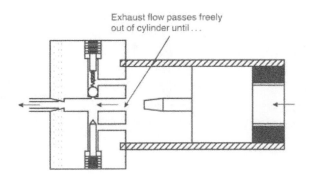

Exhaust flow passes freely
out of cylinder until . . .

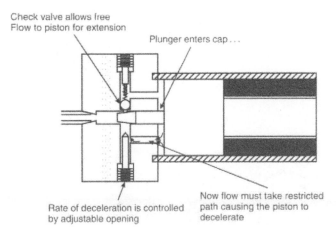

Check valve allows free
Flow to piston for extension

Plunger enters cap . . .

Rate of deceleration is controlled
by adjustable opening

Now flow must take restricted
path causing the piston to
decelerate

Figure 5.7
Operation of cylinder cushions

As shown in the figure, deceleration starts when the tapered plunger enters the opening in the cap resulting in a restriction in exhaust flow, from the barrel to the port. During the final small leg of the stroke, the oil must exhaust through an adjustable opening. The cushion design also incorporates a check valve in order to allow free flow to the piston during reversal in direction.

Consideration should be given to the maximum pressure developed by the cushions at the ends of the cylinder since any excessive pressure build-up would end up rupturing the cylinder.

5.5   Cylinder mountings

There are various types of cylinder mountings in existence. This permits versatility in the anchoring of cylinders. The rod ends are usually threaded so that they can be attached directly to the load, a clevis, a yoke or some other mating device.

Through the availability and use of various mechanical linkages, applications involving hydraulic cylinders are limited only by the ingenuity of the fluid power designer. These linkages can transform a linear motion into either an oscillating motion or a rotary motion. In addition, linkages can be employed to increase or decrease the effective leverage and stroke of a cylinder.

Listed below (Figure 5.8(a)) are schematics of typical hydraulic cylinder linkages used in a majority of modern day applications.

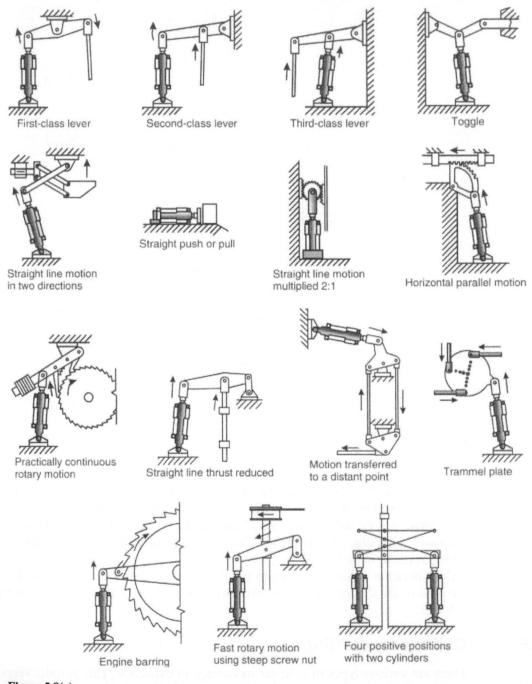

Figure 5.8(a)
Typical mechanical linkages combined with hydraulic cylinders (Courtesy of Rexnord Inc., USA)

In order to facilitate the smooth working of these mechanical linkages, various cylinder mountings have been developed, as shown in Figure 5.8(b).

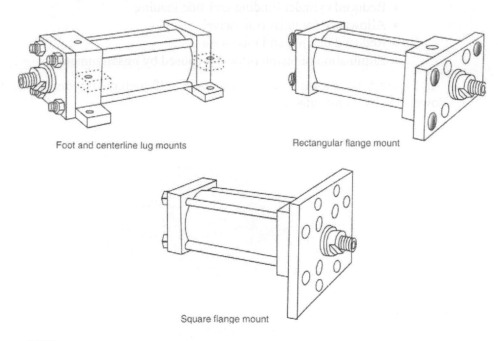

Foot and centerline lug mounts

Rectangular flange mount

Square flange mount

Figure 5.8(b)
Various cylinder mountings

The problem of side loading in hydraulic cylinders due to misalignment has been one of the most discussed topics since they have a significant bearing in evaluating cylinder life and performance. Various efforts have been undertaken from time to time by cylinder manufacturers, to minimize or eliminate this problem altogether. A fact that is increasingly being acknowledged is that it is almost next to impossible to achieve perfect alignment in cylinders.

Figure 5.9 is a schematic showing a universal cylinder-mounting accessory that has been developed to alleviate the problem of misalignment in cylinders.

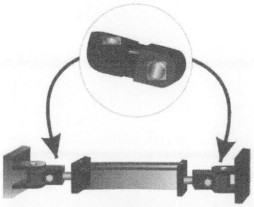

Figure 5.9
Universal alignment mounting arrangement for cylinders (Courtesy of Sheffer Corporation, Ohio)

The advantages of these types of mounting accessories are:

- Ease of mounting
- Reduced cylinder binding and side loading
- Allowance for universal swivel
- Reduced bearing and tube wear
- Elimination of piston blow-by, caused by misalignment.

In Figure 5.10, we see a unique, compact, self-contained hydraulic package called the Power-Pak by its manufacturer.

Figure 5.10
Power-Pak hydraulic package (Courtesy of Sheffer Corporation, Ohio)

It supplies force in situations warranting minimum size and maximum power. In order to provide flexibility in operation, the selection of speed and power is provided. This arrangement consists of a heavy-duty hydraulic cylinder, a reversible electric motor, a reversible generator pump, a reservoir and automatic valving. This package is a complete hydraulic power system, simple in design and easy to be put into operation.

6

Control components in a hydraulic system

6.1 Objectives

After reading this chapter, the student will be able to:

- Learn about the various control components used in hydraulics
- Understand the purpose, construction and operation of different valves such as pressure control valves, direction control valves and flow control valves and also identify the related graphical symbols in a hydraulic circuit
- Understand and explain the different methods of valve actuation
- Differentiate between compensated and non-compensated flow control
- Explain the purpose, design and operation of servo valves
- Understand the concept of hydraulic fuses
- Understand the function and purpose of shock absorbers in hydraulics
- Learn about the operation of different types of temperature and pressure switches.

6.2 Introduction

One of the most important considerations in a hydraulic system is 'control'. For any hydraulic system to function as required, a proper selection of control components is quite essential. Fluid power is primarily controlled with the help of control devices called valves. The selection of these control devices involves not only choosing the right type but also the size, actuating technique and its remote control capability.

In this chapter, we shall review in detail the working of various control devices in hydraulic systems that are as listed below:

- Control valves
- Servo valves
- Hydraulic fuses
- Temperature and pressure switches
- Shock absorbers.

6.3   Control valves

A valve is a control device used for adjusting or manipulating the flow rate of a liquid or gas in a pipeline. The valve essentially consists of a flow passage whose flow area can be varied. The external motion can originate either manually or from an actuator positioned pneumatically, electrically or hydraulically, in response to some external positioning signal. This combination of the valve and actuator is known as a control valve or an automatic control valve.

Basically there are three types of control valves:

1. *Direction control valves*: Direction control valves determine the path through which a fluid traverses within a given circuit. In other words, these valves are used to control the direction of flow in a hydraulic circuit. It is that component of a hydraulic system that starts, stops and changes the direction of the fluid flow. Additionally the direction control valve actually designates the type of hydraulic system design, either open or closed. An example of their application in a hydraulic system is the actuator circuit, where they establish the direction of motion of a hydraulic cylinder or a motor.

2. *Pressure control valves*: Pressure control valves protect the system against overpressure conditions that may occur either on account of a gradual build up due to decrease in fluid demand or a sudden surge due to opening or closing of the valves. Pressure relief, pressure reducing, sequencing, unloading, brake and counterbalance valves control the gradual buildup of pressure in a hydraulic system. Pressure surges can produce instantaneous increases in pressure as much as four times the normal system pressure and that is the reason why pressure control devices are a must in any hydraulic circuit. Hydraulic devices such as shock absorbers are designed to smoothen out pressure surges and also to dampen hydraulic shock.

3. *Flow control valves*: The fluid flow rate in a hydraulic system is controlled by flow control valves. Flow control valves regulate the volume of oil supplied to different parts of a hydraulic system. Non-compensated flow control valves are used where precise speed control is not required, since the flow rate varies with the pressure drop across a flow control valve. Pressure-compensated flow control valves are used in order to produce a constant flow rate. These valves have the tendency to automatically adjust to changes in pressure.

Since it is important to know the primary function and operation of the various types of control components, it is required to examine and study the functioning of each of these valves in detail.

6.3.1   Direction control valves

As briefly discussed above, direction control valves are used to control the direction of flow in a hydraulic circuit. They are primarily designated by their number of possible positions, port connections or ways and the manner in which they are actuated or energized. For example, the number of porting connections is designated as ways or possible flow paths. A four-way valve would comprise of four ports P, T, A and B. A Three-position valve is indicated by three connected boxes, as shown. There are several mechanisms employed for actuation or shifting of the valve. They include hand lever, foot pedal, push button, mechanical, hydraulic pilot, air pilot, solenoid and spring.

Normally open and normally closed

Direction control valves may also be categorized as normally open and normally closed valves. This terminology would normally accompany the direction control valves, as reflected in the examples of two-position valves given below (Figures 6.1(a)–(d)).

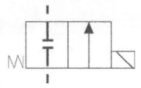

Figure 6.1(a)
Spring offset, Solenoid operated two-way valve, normally closed

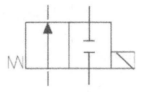

Figure 6.1(b)
Spring offset, Solenoid operated two-way valve, normally open

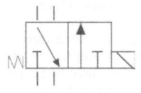

Figure 6.1(c)
Spring offset, Solenoid operated three-way valve, normally closed

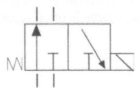

Figure 6.1(d)
Spring offset, Solenoid operated three-way valve, normally open

Direct acting and pilot-operated direction control valves

By direct acting, it is implied that some force is made to act directly on the spool, causing it to shift. A direct acting direction control valve can be actuated either manually or with the help of a solenoid.

In the example illustrated in Figure 6.2, upon energizing the solenoid, an electromagnetic force which is generated pulls the armature of the coil into the magnetic field. As a consequence of this action, the connected push pin moves the spool in the same direction, while compressing the return spring. The shift in the spool valve results in port P opening to port A and port B opening to port T or tank, thereby allowing the cylinder to extend. When the coil is de-energized, the return springs move the spool back to its center position. Manual overrides are provided with most solenoid actuated valves, allowing for the spool to be operated by hand. This can be accomplished by depressing the pin in the push pin tube end located at each end of the valve.

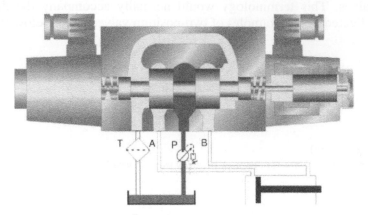

Figure 6.2
Direct acting direction control valve

In hydraulic systems requiring higher flow rates of 35 gpm and above, a greater force is required to shift the spool. This is not possible with the direct acting method and therefore a pilot-operated arrangement is used. Figure 6.3 illustrates the working of a pilot-operated direction control valve. These valves are operated by applying air pressure against a piston at either end of the valve spool. Referring Figure 6.3, the top valve is known as the pilot valve, while the bottom valve is the main valve that is to be actuated. In a typical pilot operated valve operation, the pilot valve is used to hydraulically actuate the main valve. The oil required by the pilot valve in order to accomplish this operation is directed either from an internal source or an external source. Oil is directed to one side of the main spool, when the pilot valve is energized. The resultant shift in the spool leads to the opening of the pressure port to the work port thereby directing the return fluid back to the tank. External piloting or in other words, sending fluid to the pilot valve from an external source is often resorted to.

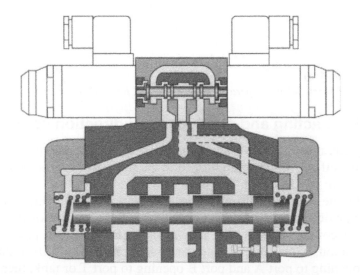

Figure 6.3
Pilot-operated direction control valve

The advantages associated with external piloting are that the effect of any other influence on the main system is not felt and the possibility of separate filtration ensures silt-free operation of the pilot valve. Additionally, the valve may also be internally or externally drained. In the case of internal draining of the pilot valve, the oil flows directly into the tank chamber of the main valve. When operating the main control spool, pressure or flow surges occurring in the tank port may affect the unloaded side of the main valve as well as the pilot valve. This may be avoided by externally draining the pilot valve or in other words, feeding the pilot oil flow back to the tank.

To understand the concept of Pilot-operated direction control valves even better, let us consider the following example shown in Figure 6.4 illustrating the cutaway section of a pilot-operated four-way valve.

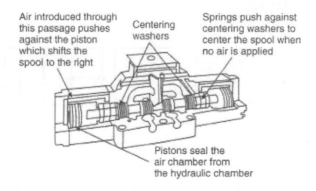

Figure 6.4
Air pilot-operated four-way valve (Courtesy of Sperry Vickers, Michigan)

As shown in the figure, the springs located at both ends of the spool push against the centering washers to center the spool when no air is applied. When air is introduced through the left passage, its pressure pushes against the piston to shift the spool to the right. Similarly, when air is introduced through the right passage, its pressure pushes against the piston to shift the spool to the left. The graphical representation of this valve is shown in Figure 6.5.

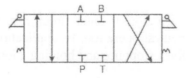

Figure 6.5
Graphical symbol for pilot-operated four-way valve

Open and closed center hydraulic circuits

A majority of the hydraulic circuits are basically categorized into two types, open center and closed center. The type of circuit is usually designated by direction control valves. In open center circuits, the pump flow is routed back to the tank through the direction control valve during neutral or dwell time. Normally in this type of circuit, a fixed displacement pump such as gear pump is used. In the event of flow being blocked during neutral or the direction control valve being centered, flow tends to get forced over the relief valve, possibly leading to the formation of an excessive amount of heat.

In the closed center circuit, the pump flow is blocked at the direction control valve both in neutral or when the valve is centered. In this case, either a pressure-compensated pump such as a piston pump that de-strokes or an unloading circuit with a fixed displacement pump is used.

A neutral or central position is provided in a three-position direction control valve. This determines whether the circuit is open or closed and also the type of work application depending on the inter-connection between the P and T ports and the configuration of the A and B ports respectively. The four commonly used three-position direction control valves: the open type, the closed type, the tandem type and the flow type are illustrated in Figures 6.6(a)–(d).

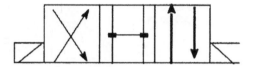

Figure 6.6(a)
Open-type DCV

Referring Figure 6.6(a), the P, T, A and B ports in the open type are connected together giving an open center and work force draining to the tank. An example of its application is the free wheeling option in neutral, in motor circuits.

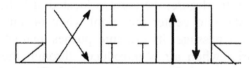

Figure 6.6(b)
Closed-type DCV

In the closed type, a closed center circuit results on account of the blocking of the P, T, A and B ports in neutral operation. Its application can be traced to parallel circuits where the operation involves stopping and holding a load in mid-cycle.

Figure 6.6(c)
Float-type DCV

In the float type, the A and B ports are interconnected to T, while the P port is blocked. Since P is blocked, the circuit becomes a closed center. An example of this type is its application in parallel circuits where a hydraulic motor is freewheeled in neutral.

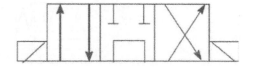

Figure 6.6(d)
Tandem-type DCV

In the tandem type, port P is connected to port T and the ports A and B are blocked. This results in an open circuit. This type finds application in circuits involving fixed volume pumps where on account of ports A and B being blocked, the load can be held in neutral.

Let us discuss some of the most commonly used direction control valves in hydraulic circuits.

Check valve

Figures 6.7(a) and (b) show the symbolic representation of a check valve along with a simple check valve application in an accumulator circuit. As the name implies, direction control valves are used to control the direction of flow in a hydraulic circuit. The simplest type is a check valve, which is a one-way direction control valve. It is a one-way valve because it permits free flow in one direction and prevents any flow in the opposite direction.

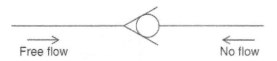

Figure 6.7(a)
Symbolic representation of a check valve in a hydraulic circuit

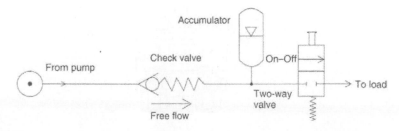

Figure 6.7(b)
Check valve application in a hydraulic circuit

Figure 6.8 shows the internal operation of a check valve. As shown, a light spring holds the poppet in the closed position. In the free flow direction, the fluid pressure overcomes the spring force. If the flow is attempted in the opposite direction, the fluid pressure pushes the poppet (along with the spring force) in the closed position. Therefore, no flow is permitted. The higher the pressure, the greater will be the force pushing the poppet against its seat.

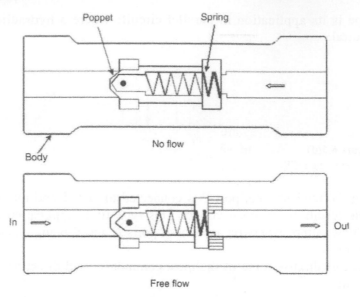

Figure 6.8
Operation of a check valve

Pilot-operated check valve

The second type of check valve is the pilot operated check valve. The cross-section of a typical pilot operated check valve has been illustrated in Figure 6.9.

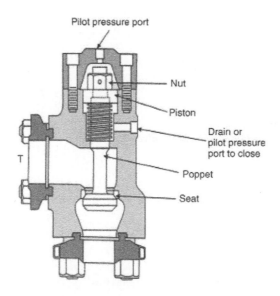

Figure 6.9
Pilot-operated check valve (Courtesy of Sperry Vickers, Michigan)

This type of check valve always enables free flow in one direction but permits flow in the normally blocked opposite direction also if the pilot pressure is applied at the pilot pressure port of the valve. The check valve poppet has a pilot piston attached to the threaded poppet stem by a nut. The light spring holds the poppet seated in a no-flow condition by pushing against the pilot piston. The purpose of the separate drain port is to prevent oil from creating a pressure build up at the bottom of the piston.

Pilot check valves are often used in hydraulic systems where it is desirable to stop the check action of the valve for a portion of the equipment cycle. An example of its application is its use in locking hydraulic cylinders in position.

Spool-type direction control valves

As discussed earlier, in spool-type direction control valves, spools incorporated in the control valve body are used to provide different flow paths. This is accomplished by the opening and closing of discrete ports by the spool lands. The spool is a cylindrical member which has large-diameter lands, machined to slide in a very close-fitting bore of the valve body. The radial clearance is usually less than 0.02 mm. The spools may be operated through different means like mechanical actuation, manual operation, pneumatic operation, hydraulic or pilot control and electrical operation.

Two-way directional valves

This type of directional valve is designed to allow flow in either direction between two ports. Figure 6.7(b) above showing a check valve application in an accumulator circuit is a typical example of a two-way, two-position on–off valve. Its function is to connect the accumulator to the load whenever desired. To put it rather simply, this valve is the hydraulic equivalent of a regular single-pole, single-throw (SPST) on–off electrical switch.

Three-way and four-way direction control valves

An additional type of direction control valve is the three-way and four-way valve, containing three and four ports respectively. Figure 6.10(a) depicts the flow paths through two four-way valves. As shown in the figure, one of these valves is used as a three-way valve since its port T leading to the oil tank is blocked. One of the simplest ways by which a valve port could be blocked is by screwing a threaded plug into the port opening.

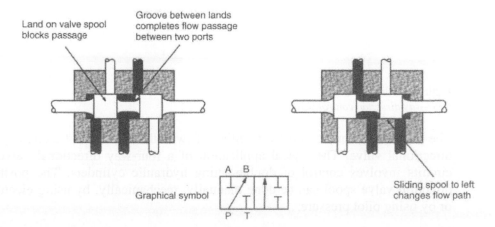

Figure 6.10(a)
Four-way direction control valve used as a three-way valve

Referring figure, the flow entering at pump port 'P' (port connected to the pump discharge line) can be directed to either of the outlet ports 'A' and 'B'. Most direction control valves use a sliding spool to change the path of flow through the valve. For a given position of the spool, a unique flow path configuration exists within the valve. Directional valves are designed to operate with either two positions of the spool or three positions of the spool. Let us now analyze the flow paths through each valve shown in the figure.

Three-way valve (four-way valve used as a three-way valve)

Here the port T is blocked and only the other three ports A, B and P are used.

The flow can go through the valve in two unique ways depending on the spool position:

- *Spool position 1*:   Flow can go from P to B as shown by the straight through line and arrow. Port A is blocked by the spool in this position.
- *Spool position 2*:   Flow can go from P to A. Port B is blocked by the spool in this position.

Four-way valve

The flow can go through the valve in four unique ways depending on the spool position: Referring Figure 6.10(b),

- *Spool Position 1*:   Flow can go from P to A and B to T.
- *Spool Position 2*:   Flow can go from A to T and P to B.

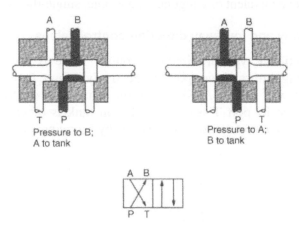

Spool positions in four-way valve

Figure 6.10(b)
Four-way direction control valve

The pump flow can be directed to either of two different parts of a circuit by a three-way directional valve. The typical application of a four-way directional valve in hydraulic circuits involves control of double-acting hydraulic cylinders. The positioning of the direction valve spool can be done manually, mechanically, by using electrical solenoids or by using pilot pressure.

Solenoid-operated direction control valves

The most common way of actuating the spool valve is by using a solenoid. Figure 6.11 shows a typical solenoid-operated directional valve.

When the electrical coil energizes, it creates a magnetic force that pulls the armature into the coil. This causes the armature to exert a pushing force on the push rod to move the spool of the valve. Solenoids are provided at both ends of the spool. The example shown above is that of a four-way, three-position, spring-centered direction control valve.

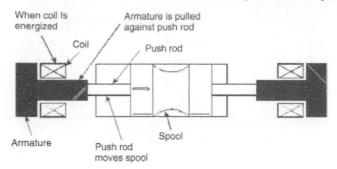

Figure 6.11
Operation of a solenoid-operated direction control valve

Figure 6.12 is an illustration of the cutaway section of an actual solenoid-operated direction control valve manufactured by Continental Hydraulics.

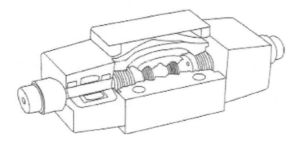

Figure 6.12
Actual solenoid-operated direction control valve (Courtesy of Continental Hydraulics, Minnesota)

This valve has a flow capacity of 50 lpm and a maximum operating pressure of 250 kg/cm^2 (3555 psi). It has a wet armature solenoid. The fluid around the armature serves to cool it and cushion its strokes without affecting the response time. There are no seals around the armature because of which its movement is not restricted. This allows all the power developed by the solenoid to be transmitted to the spool valve without the need to overcome seal friction.

Rotary four-way direction control valves

Although most direction control valves are of spool type design, other types are also used. One such design is the rotary four-way valve, which consists of a rotor closely fitted in the valve body.

The passages in the rotor connect or block-off the ports in the valve body to provide the four flow paths. The design shown above is a three-position valve in which the centered position has all the four ports blocked. Rotary valves are usually actuated either manually or mechanically. The operation of this valve is illustrated below (Figure 6.13).

This design contains lapped metal-to-metal sealing surfaces which form a virtually leak proof seal. The gradual overlapping of the round flow passages produce a smooth shearing action which results in lesser load on the handle during operation and absence of sudden surges. Also there is no external leakage because of the presence of a static seal on the rotating shaft (non-reciprocating and non-pressurized). The high-pressure regions are confined to flow passages. This type of valve can take higher velocities and more flow than a spool valve of the same size.

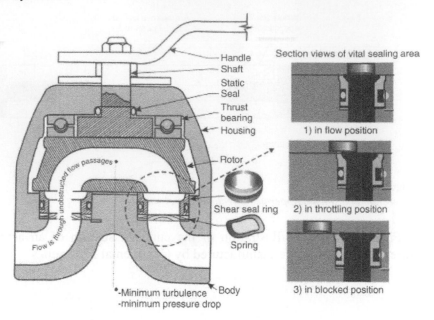

Figure 6.13
Operation of a rotary control valve (Courtesy of DeLaval Turbine Inc., California)

These valves are available in a variety of three-way and four-way and two- and three-position flow path configuration.

Shuttle valves

This is another type of direction control valve. It allows a system to operate from either of two fluid power sources. One application is for safety in the event that the main pump can no longer provide the hydraulic power to operate emergency devices.

As soon as the primary source is exhausted, the shuttle valve shifts to allow fluid to flow from the secondary backup pump. A typical construction of this type of valve is shown in Figure 6.14.

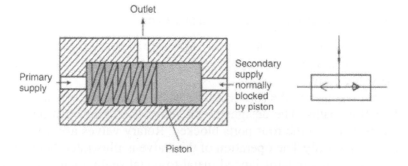

Figure 6.14
Shuttle valve

The shuttle valve consists of a floating piston, which can be shuttled to either side of the valve depending on which side of the piston has a greater pressure. Shuttle valves may be spring loaded in one direction to favor one of the supply sources. The shuttle valve is essentially a direct acting double check valve with a cross bleed, as depicted by the graphical symbol. The double arrows in the graphical symbol indicate that reverse flow is permitted.

6.3.2 Pressure control valves

We have already briefly discussed what a pressure control valve is and what it does in a hydraulic system. This section is designed to give a deeper understanding of the concept of manipulating force through a hydraulic system using pressure control valves and to also illustrate the operating and design principles of the various types of pressure control valves and their applications. The two basic pressure control valve design types are:

1. Direct-acting pressure control valves and
2. Pilot-operated pressure control valves.

The operating principles of all the pressure control valves revolve around these two basic design types.

The primary objective in any hydraulic circuit is to either control the flow rate or pressure. For accurate control of force in a hydraulic circuit, six different types of pressure control valves have been developed. These are given below along with their graphical representation (Figures 6.15 (a)–(f)).

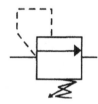

Figure 6.15(a)
Relief valve

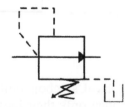

Figure 6.15(b)
Reducing valve

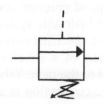

Figure 6.15(c)
Unloading valve

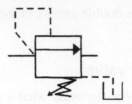

Figure 6.15(d)
Sequence valve

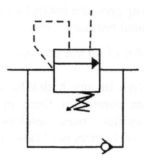

Figure 6.15(e)
Counterbalance valve

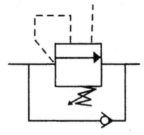

Figure 6.15(f)
Brake valve

One may find the graphical symbols quite confusing since these valves resemble one another so closely that often only their location in a hydraulic circuit may actually help determine what type of pressure valve they are.

Simple pressure relief valve

The most widely used type of pressure control valve is the pressure relief valve since it is found in practically every hydraulic system. It is a normally closed valve whose function is to limit the pressure to a specified maximum value by diverting the pump flow back to the tank. The primary port of a relief valve is connected to system pressure and the secondary port connected to the tank. When the poppet in the relief valve is actuated at a predetermined pressure, a connection is established between the primary and secondary ports resulting in the flow getting diverted to the tank. Figure 6.16, illustrates the operation of a simple direct acting relief valve.

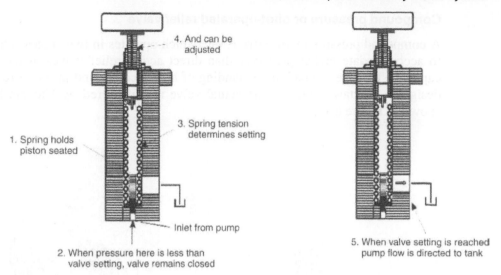

Figure 6.16
Simple pressure relief valve

A poppet is held seated inside the valve by the direct force of a mechanical spring which is usually adjustable. The poppet is kept closed by the spring tension set on the knob until the system pressure working against the poppet reaches the cracking pressure. The poppet is forced off its seat when the system pressure reaches full relief value. This permits fluid flow across the poppet to the tank. Thus the required pressure in the system is maintained as per the set value on the pressure relief valve.

When the hydraulic system does not accept any flow due to a safety reason in the system, the pressure relief valve releases the fluid back to the tank to maintain the desired system pressure in the hydraulic circuit. It provides protection against any overloads experienced by the actuators in the hydraulic system. One important function of a pressure relief valve is to limit the force or torque produced by the hydraulic cylinders and motors.

One important consideration to be taken note of is the practical difficulty in designing a relief valve spring strong enough to keep the poppet closed at high-flow and high-pressure conditions. This is normally the reason why direct acting relief valves are available only in relatively smaller sizes.

A partial hydraulic circuit consisting of a pump and a pressure relief valve has been depicted in Figure 6.17. The pump and the relief value are symbolically represented.

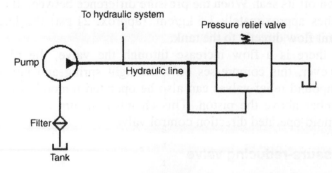

Figure 6.17
Symbolic representation of a partial hydraulic circuit, showing a relief valve

Compound pressure or pilot-operated relief valve

A compound pressure relief valve is one which operates in two stages. They are designed to accommodate higher pressures than direct acting relief valves at the same flow rate capacity. To have a broad understanding of how a compound pressure relief is internally designed, a cutaway view of an actual valve manufactured by Vickers INC., Detroit is shown in Figure 6.18.

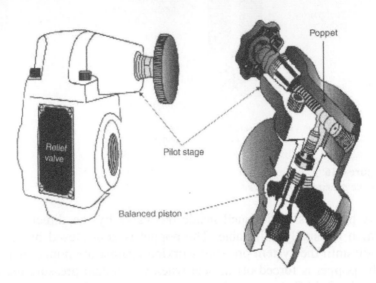

Figure 6.18
External and cutaway views of an actual compound relief valve (Courtesy of Sperry Vickers, Michigan)

The first stage of the pilot relief valve includes the main spool which is normally closed and kept in position by a non-adjustable spring. The pilot stage is located in the upper valve body and contains a pressure-limiting poppet, which is held against a seat by an adjustable spring. The lower body contains the port connections. The balanced piston in the lower part of the body accomplishes diversion of the full pump flow.

In normal operation, the balanced piston is in a condition of hydraulic balance. Pressure at the inlet port acts on both sides of the piston, through an orifice, that is drilled through the large land. For pressures less than the valve setting, the piston is held on its seat by a light spring. As soon as the pressure reaches the setting of the adjustable spring, the poppet is forced off its seat. This limits the pressure in the upper chamber. The restricted flow through the orifice into the upper chamber results in an increase in pressure in the lower chamber. This causes an imbalance in the hydraulic forces, which tends to raise the piston off its seat. When the pressure difference between the upper and the lower chamber reaches approximately 1.5 kg/cm^2 (approx. 21 psi) the large piston lifts off its seat to permit flow directly to the tank.

If there is a flow increase through the valve, the piston lifts further off its seat. However, this compresses only the light spring and hence very little override occurs. Compound relief valves can also be operated remotely by using the outlet port from the chamber above the piston. This chamber in turn can be vented to the tank through a solenoid-operated direction control valve.

Pressure-reducing valve

Pressure-reducing valves are normally open pressure control valves that are used to limit pressure in one or two legs of a hydraulic circuit. Reduced pressure results in a reduced force being generated. This is the only pressure control valve which is of the normally open type (Figure 6.19(a)). A typical pressure-reducing valve and its function is described below (Figure 6.19(c)).

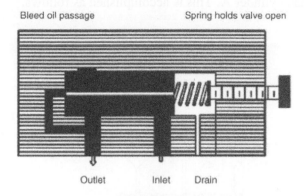

Figure 6.19(a)
Normal open position of the valve permitting free fluid flow from the inlet to the outlet

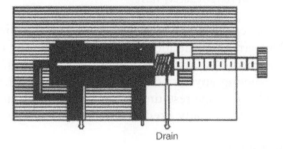

Figure 6.19(b)
Closing of the valve due to outlet pressure increasing to the set value of the valve

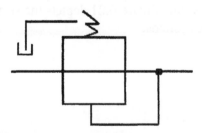

Figure 6.19(c)
Operation of a pressure-reducing valve

This valve is actuated by the downstream pressure and tends to close as the pressure reaches the valve setting. When the downstream pressure is below the valve setting, fluid will flow freely from the inlet to the outlet. Observe that there is an internal passage from the outlet, which transmits the outlet pressure to the spool end opposite the spring. When the downstream pressure increases beyond the value of the spring setting, the spool moves to the right to partially block the outlet port as shown in Figure 6.19(b). Just enough flow is thus passed through the outlet to maintain its preset pressure. If the valve closes completely, leakage past the spool could cause the downstream pressure to build above the set pressure of the spring. This is prevented from occurring by allowing a continuous bleeding to the tank through a separate drain line.

Practical application of a pressure reducing valve in a hydraulic system

Let us consider a hydraulic circuit where one cylinder is required to apply a lesser force than the other as shown in Figure 6.20. Here cylinder B is required to apply a lesser force than cylinder A. This is accomplished as follows.

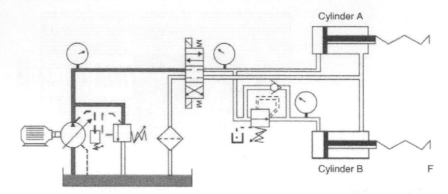

Figure 6.20
Application of a pressure-reducing valve

A pressure-reducing valve is placed just before cylinder B in the hydraulic circuit as shown. This arrangement allows flow to the cylinder, until the set pressure value on the valve is reached. At this point where the set pressure is reached, the valve shuts off, thereby preventing any further buildup of pressure. The fluid is bled to the tank through the drain valve passage resulting in the easing-off of the pressure, as a result of which the valve opens again. Finally a reduced modulated pressure equal to the valve results.

Unloading valve

Unloading valves are remotely piloted, normally closed pressure control valves, used to direct flow to the tank when pressure at a particular location in a hydraulic circuit reaches a predetermined value. Figure 6.21 depicts the sectional view of a typical unloading valve used in hydraulic systems.

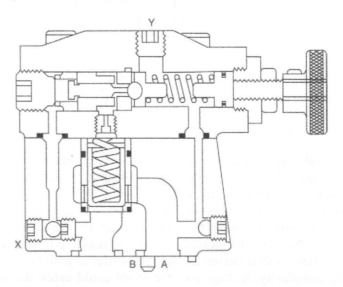

Figure 6.21
Cross-sectional view of an unloading valve

The unloading valve in Figure 6.21 is used to unload pressure from the pump connected to port A, when the pressure at port X is maintained at a value satisfying the valve setting. The spring-loaded ball exercises control over the high-flow poppet along with the pressure applied at port X. Flow entering at port A is blocked by the poppet at low pressures. The pressure signal from port A passes through the orifice in the main poppet to the top side area and then to the ball. There is no flow through these sections of the valve until the pressure rise equals the maximum value permitted by the spring-loaded ball. When that occurs, the poppet lifts causing fluid flow from port A to port B which in turn is connected to the tank. The pressure signal at port X acts against the solid control piston and forces the ball further off the seat. Due to this, the topside pressure on the main poppet reduces and allows flow from port A to B with a very low-pressure drop, as long as the signal pressure at port X is maintained.

Application

A typical example of an unloading valve application is a high–low system consisting of two pumps, one a high displacement pump and the other a low displacement pump as shown in Figure 6.22.

This system shown above is designed for providing a rapid return on the work cylinder. In this system, the net total displacement of both the pumps is delivered to the work cylinder until the load is contacted. At this point there is an increase in system pressure and this causes the unloading valve to open. This results in the flow from the high displacement pump getting directed back to the tank at a minimal pressure. The low volume pump continues to deliver flow for the higher pressure requirement of the work cycle. For facilitating rapid return of the cylinder, flow from both the pumps is again utilized.

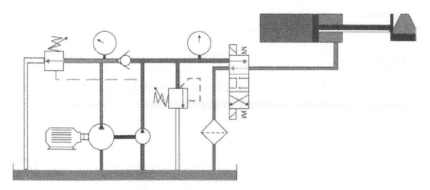

Figure 6.22
High–low system

Sequencing valve

A sequencing valve again is a normally closed pressure control valve used for ensuring a sequential operation in a hydraulic circuit, based on pressure. In other words, sequencing valves ensure the occurrence of one operation before the other. A sectional view of a sequencing valve is shown in Figure 6.23.

When the components connected to port A of the valve reach the pressure set on the valve, the fluid is passed by the valve through port B to do additional work in a different portion of the system. The high-flow poppet of the sequence valve is controlled by the spring-loaded cone. At low pressures, the poppet blocks the flow of fluid from entering port A. The pressure signal at port A passes through the orifices to the top side of the poppet and to the cone. There is no flow through the valve unless the pressure at port A exceeds the maximum set pressure on the spring-loaded cone. When the pressure reaches the set valve, the main poppet lifts, allowing the flow to pass through port B. It maintains the adjusted pressure at port A until the pressure at port B rises to the same value. A small pilot flow (about ¼ gpm) goes through the control piston and past the pilot cone to the external drain. When there is subsequent pressure increase in port B, the control piston acts to prevent further pilot flow loss. The main poppet opens fully and allows the pressures at port A and B to rise together. Flow may go either way during this condition.

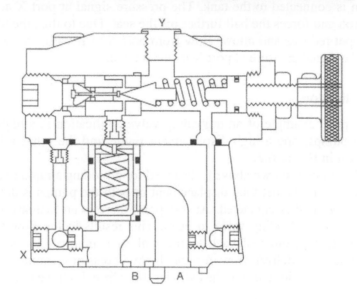

Figure 6.23
Cross-sectional view of a sequencing valve

Application

Let us consider a hydraulic circuit in which two cylinders are used to execute two separate operations as shown in Figure 6.24.

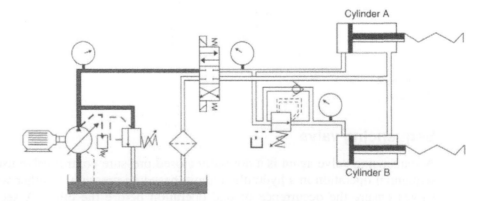

Figure 6.24
Sequencing operation in a hydraulic circuit

Now, let us assume that cylinder A is required to extend completely before cylinder B extends. This can be accomplished by placing a sequencing valve just before cylinder B as shown. The pressure value of the valve is set to a predetermined value say 28 kg/cm²

(400 psi). This ensures that the operation involving cylinder B will occur after the operation involving cylinder A or in other words, cylinder B will not extend before a pressure of 28 kg/cm^2 (400 psi) is reached on cylinder A.

Counterbalance valve

A counterbalance valve again is a normally closed pressure control valve and is particularly used in cylinder applications for countering a weight or overrunning load. Figure 6.25 shows the operation of a typical counterbalance valve.

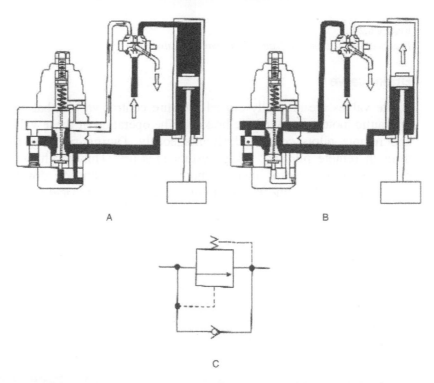

A   B

C

Figure 6.25
Operation of a counterbalance valve

The primary port of this valve is connected to the bottom of the cylinder and the secondary port is connected to the direction control valve (DCV). The pressure setting of the counterbalance valve is kept higher than required to prevent the cylinder load from falling.

When the pump flow is directed to the top of the cylinder through the DCV, the cylinder piston is pushed downward. This causes the pressure at the primary port to increase and raise the spool. This results in the opening of a flow path for discharge through the secondary port to the DCV and back to the tank.

When raising the cylinder, an integral check valve opens to allow free flow for retraction of the cylinder. Figure 6.26 is an illustration of how exactly the counterbalance valve operates in a hydraulic circuit. As shown in the figure, the counterbalance valve is placed just after the cylinder in order to avoid any uncontrolled operation. In the event of the counterbalance valve being not provided, there would be an uncontrolled fall of the load, something which the pump flow would find hard to keep pace with. The counterbalance valve is set to a pressure slightly higher than the load-induced pressure. As the cylinder is extended, there must be a slight increase in pressure in order to drive the load down.

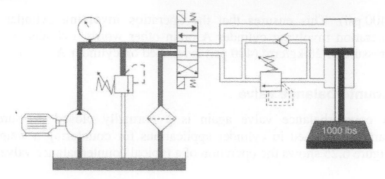

Figure 6.26
Practical application involving counterbalance valve

Brake valve

Brake valves are normally closed pressure control valves that are frequently used with hydraulic motors for dynamic braking. The operation of these valves involves both direct and remote pilots connected simultaneously. During running, the valve is kept open through remote piloting, using system pressure. This results in eliminating any back pressure on the motor that might arise on account of downstream resistance and subsequent load on the motor. Figure 6.27 shows the operation of a brake valve in a motor circuit.

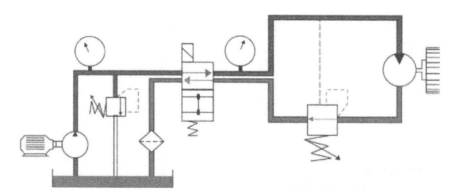

Figure 6.27
Practical application of a brake valve

When the direction control valve is de-energized, remote pilot pressure is lost allowing the valve to close. The valve is then driven open through the internal pilot, by the inertia of the load, resulting in dynamic braking.

6.3.3   Flow control valves

A flow control valve is a device used for adjusting or manipulating the flow rate of a liquid or a gas in a pipeline. The valve contains a flow passage or a port whose flow area can be varied. The role of a flow control valve in a hydraulic circuit is very important and its very location is critical to optimum system performance.

The basic function of a flow control valve is to reduce the rate of flow in its leg of a hydraulic circuit. One of the most important applications of flow control valves in hydraulic systems is in controlling the flow rate to cylinders and motors to regulate their speeds. Any reduction in flow will in turn, result in a speed reduction at the actuator.

There are many different designs of valves used for controlling flow. Many of these designs have been developed to meet specific needs.

Some factors, which should be considered during the design stage of a flow control valve are:

- The maximum and minimum flow rates and the fluid density, which affect the size of the valve
- The corrosive property of the fluid, which determines the material of construction of the valve
- The pressure drop required across the valve
- The allowable leakage limit across the valve in its closed position
- The maximum amount of noise from the valve that can be tolerated
- The means of connecting the valve to the process i.e. screwed, flanged or butt welded.

Flow control valves are classified as:

Fixed or non-adjustable flow control valves represented symbolically as in Figures 6.28(a)–(d).

Figure 6.28(a)
Fixed flow control valve

Adjustable flow control valves represented in hydraulic circuits as

Figure 6.28(b)
Adjustable flow control valve

Additionally they may also be classified as:
Throttling

Figure 6.28(c)
Throttling flow control valve

and pressure-compensated flow control valves represented as:

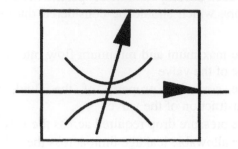

Figure 6.28(d)
Pressure compensated flow control valve

Let us study in detail the common flow control valve types employed in hydraulic circuits from their operational, functional and application point of view.

Globe valve

This is the simplest form of a flow control valve. The globe valve gets its name from the disk element 'globe' that presses against the valve seat to close the valve.

A simplified view of a globe valve has been illustrated in Figure 6.29.

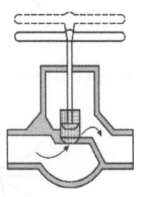

Figure 6.29
Simplified view of a globe valve

The fluid flow through the valve is at right angles to the direction of flow in pipes. When this valve is opened, the entire surface of the globe moves away from the valve seat at once. Due to this action, a globe valve provides an excellent means of throttling the flow. In a hydraulic system, the globe valve can be operated either manually by means of a hand wheel or mechanically by means of an actuator.

Butterfly valve

This is another type of flow control valve. It consists of a large disk which is rotated inside a pipe, the restriction in flow being determined by the angle. Figure 6.30 shows a simple design of a butterfly valve.

The advantage with this valve is that it can be constructed to almost any size. These valves are widely used for controlling gas flow. But a major problem associated with these valves is the high amount of leakage in the shut-off position.

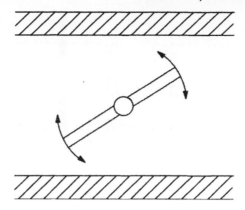

Figure 6.30
Butterfly valve

Ball valve

This is another type of flow control valve shown in Figure 6.31.

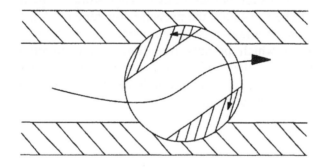

Figure 6.31
Ball valve

It is made up of a ball with a through hole which is rotated inside a machined seat. The manner in which flow control is exercised can be understood better with the help of Figures 6.32(a) and (b).

From Figure 6.32(a), it can be seen how flow assists opening and opposes closing of the valve. Conversely, from Figure 6.32(b), the flow is seen to assist closing and oppose opening of the valve.

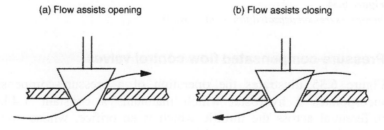

Figure 6.32
Flow control in a ball valve

Figure 6.33 shows the balanced version of a ball valve. This valve uses two plugs and two seats with opposite flows resulting in very little dynamic reaction onto the actuator shaft, although at the expense of higher leakage.

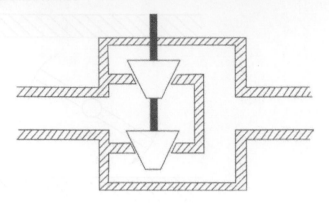

Figure 6.33
Balanced ball valve

Throttling only or non-pressure-compensated valve

This type of valve is used where the system pressures are relatively constant and the motoring speeds are not too critical. They work on the principle that the flow through an orifice will be constant if the pressure drop remains constant.

The figure of a non-pressure-compensated valve shown in Figure 6.34 also includes a check valve which permits free flow in the direction opposite to the flow control direction. When the load on the actuator changes significantly, the system pressure changes. Thus, the flow rate through the non-pressure-compensated valve will change for the same flow rate setting.

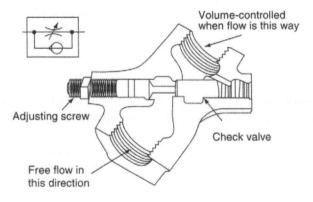

Figure 6.34
Non-pressure-compensated flow control valve

Pressure-compensated flow control valves

Figure 6.35 illustrates the operation of a pressure-compensated valve. The design incorporates a hydrostat which maintains a constant $1.4\,\text{kg/cm}^2$ (20 psi) pressure differential across the throttle which is an orifice, whose area can be adjusted by an external knob setting. The orifice area setting determines the flow rate to be controlled. The hydrostat is normally held open by a light spring. However, it starts to close as inlet pressure increases and overcomes the spring tension. This closes the opening through the hydrostat, thereby blocking all the flow in excess of the throttle setting. As a result, the only amount of fluid that can flow through the valve is that amount which a $1.4\,\text{kg/cm}^2$ (20 psi) pressure can force through the throttle.

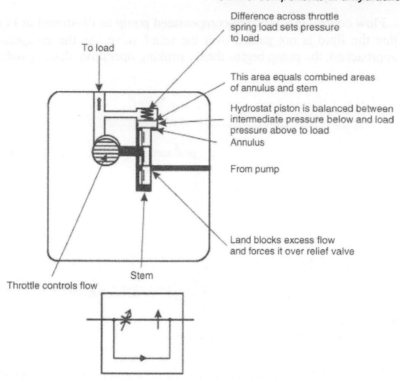

To load

Difference across throttle
spring load sets pressure
to load

This area equals combined areas
of annulus and stem

Hydrostat piston is balanced between
intermediate pressure below and load
pressure above to load
Annulus

From pump

Land blocks excess flow
and forces it over relief valve

Throttle controls flow

Stem

Figure 6.35
Pressure-compensated flow control valve

To understand better the concept of pressure compensation in flow control valves, let us try and distinguish between flow control in a fixed displacement pump and that in a pressure-compensated pump. Figure 6.36 is an example of flow control in a hydraulic circuit with fixed volume pumps.

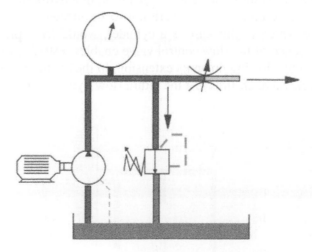

Figure 6.36
Flow control with fixed volume pumps

In this system, a portion of the fluid is bypassed over the relief valve in order to reduce flow to the actuator. Pressure increases upstream as the flow control valve, which in this case is a needle valve, is closed. As the relief pressure is approached, the relief valve begins to open, bypassing a portion of the fluid to the tank.

Flow control in a pressure-compensated pump as illustrated in Figure 6.37, is different in that the fluid is not passed over the relief valve. As the compensator setting pressure is approached, the pump begins the de-stroking operation, thereby reducing the outward flow.

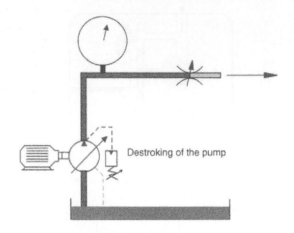

Figure 6.37
Flow control with pressure-compensated pumps

The design of a pressure-compensated flow control valve is such that it makes allowances for variations in pressure, before or after the orifice. In a pressure-compensated flow control valve, the actuator speed does not vary with variation in load.

Meter-in and meter-out functions

Meter-in is a method by which a flow control valve is placed in a hydraulic circuit in such a manner that there is a restriction in the amount of fluid flowing to the actuator. Figure 6.38(a) shows a meter-in operation in a hydraulic system.

If the flow control valve were not to be located, the extension and retraction of the actuator which in this case is a cylinder, would have proceeded at an unrestricted rate. The presence of the flow control valve enables restriction in the fluid flow to the cylinder and thereby slowing down its extension. In the event of the flow direction being reversed, the check valve ensures that the return flow bypasses the flow control valve.

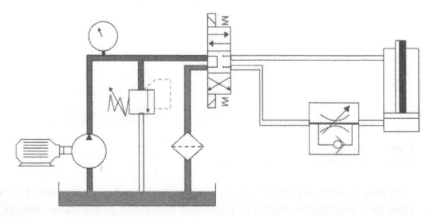

Figure 6.38(a)
Meter-in operation

For the same meter-in operation, Figure 6.38(b) shows shifting of the flow control to the other line. This enables the actuator to extend at an unrestricted rate but conversely the flow to the actuator during the retracting operation can be restricted so that the operation takes place at a reduced rate. The meter-in operation is quite accurate with a positive load. But with an overrunning load over which the actuator has no control, the cylinder begins to cavitate.

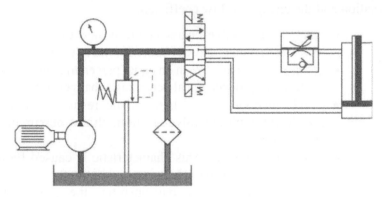

Figure 6.38(b)
Meter-in operation

Meter-out operation

In the meter-out operation shown in Figure 6.39, the direction of the flow through the circuit is simply changed as can be made out from the diagram. It is the opposite of a meter-in operation as this change in direction will cause the fluid leaving the actuator to be metered. The advantage with the meter-out operation is that unlike in the case of meter-in operation, the cylinder here is prevented from overrunning and consequent cavitating.

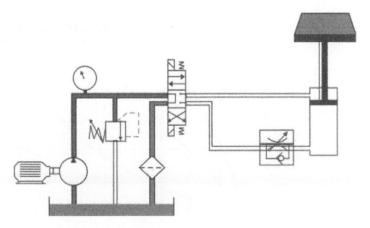

Figure 6.39
Meter-out operation

One major problem confronting the meter-out operation is the intensification of pressure in the circuit which can in turn occur on account of a substantial differential area ratio between the piston and the rods. Pressure intensification occurs on the rod side when the meter-out operation is carried out without a load on the rod side of the cylinder and can result in failure of the rod seals. It is therefore seen that both the meter-in and meter-out operations have their relative advantages and disadvantages and only the application determines the type and nature of flow valve placement.

Valve characteristics

The inherent valve flow characteristics describe the relationship between valve travel or rotation and the change in flow coefficient:

- *Linear*:   The valve characteristic is said to be linear when the change in the flow coefficient is directly proportional to the change in the valve travel.
- *Equal percentage*:   With an equal percentage characteristic, equal increments of valve travel produce equal percentage changes in the existing flow coefficient.
- *Quick opening*:   This characteristic results in a rapid increase in the flow coefficient, with the valve reaching almost maximum capacity in its first 50% of the travel.
- *Shape of opening*:   This characteristic is caused by a change in the shape of the port as valve travel changes.
- *Capacity*:   The larger the opening, the greater is the flow coefficient. Therefore, at maximum valve travel, the equal percent characteristic will have the lowest flow coefficient.

The graph below contains a graphical representation of the above characteristics (Figure 6.40).

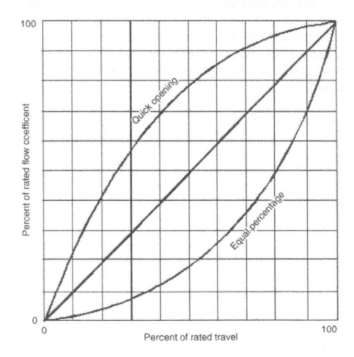

Figure 6.40
Valve sizing

In selecting a control valve, while factors such as valve material, pressure and temperature ratings are very important, choosing the correct valve size also assumes equal importance. Simply specifying a valve size to match the existing pipeline size is impractical and can lead to improper functioning of the entire system. Obviously, a valve which is too small will not give the rated flow rate while a valve too large in size would be rather expensive and result in improper control.

Using the principle of conservation of energy, Daniel Bernoulli discovered that as a liquid flows through an orifice, the square of the fluid velocity is directly proportional to the pressure differential across the orifice and inversely proportional to the specific gravity of the fluid. Hence, greater the pressure differential, higher will be the velocity while on the other hand, greater the fluid density, lower will be the velocity. Logically, the fluid flow rate for liquids can be calculated by multiplying the fluid velocity times the flow area.

After taking into account the proportionality relationship, energy losses due to friction, turbulence and varying discharge coefficient for various orifices, the sizing equation can be written as follows:

$$K_V = \left\{ \frac{Q}{(31.6 \times R)} \right\} \times \sqrt{\left(\frac{\varphi}{\Delta p} \right)}$$

Where K_V is the flow in m^3/h of water at a pressure differential of 1 atmosphere. It is known as the valve-sizing coefficient and is a function of length, diameter and material friction:

Q is the flow in m^3/h

R is the reduction factor.

This reflects the ratio of pressure drop across the valve (due to flashing and cavitations) and the pressure recovery profile of the system.

φ is the density in kg/m^3

Δp is the pressure drop in psi.

For a given flow rate, a high K_V corresponds to a lower Δp. However, valve sizing is usually carried out on the basis of the following equation,

$$Q = C_V \sqrt{\frac{\Delta p}{G}}$$

Where

Q is the flow rate in gpm

C_V is the sizing coefficients for liquids

Δp is the pressure drop in psi

G is specific gravity.

To size a valve, it is required to calculate the values of K_V and C_V at maximum flow rate conditions using a value of Δp, which is allowable. Initial valve selection is to be made by using a graph or chart allowing a valve travel of less than 90% at maximum flow and not less than 10% at minimum flow.

6.4   Servo valves

Introduction

The hydraulic systems, subsystems and hydraulic components that have been discussed so far have had open-loop control or in other words power transfer without feedback. We shall now take a look at servo or closed loop control coupled with feedback sensing devices, which provide for a very accurate control of position, velocity and acceleration of an actuator.

A servo valve is a direction control valve, which has an infinitely variable positioning capability. Thus, it controls not only the direction of the fluid flow but also the quantity. In a servo valve, the output controlled parameter is measured with a transducer and fed back to a mixer where the feedback is compared with the command. The difference is expressed in the form of an error signal which is in turn used to induce a change in the system output, until the error is reduced to zero or near zero. A typical example is the use of a thermostat in an automatic furnace whose function is to measure the room temperature and accordingly increase or decrease the heat in order to keep it constant. Let us now discuss in brief, the various components that comprise a servo system.

Servo components

Supply pumps

Servo systems generally require a constant pressure supply. Since fixed displacement pumps give out excess heat resulting in power loss, pressure-compensated pumps are commonly employed as they are ideally suited for servo operations.

Servo motors

Piston motors are generally preferred over gear or vane-type motors because of their decreased levels of internal leakage. Both in-line and bent-axis type piston motors are used in servo operations although the in-line type has more frictional drag. This is not a serious limitation because in usual circumstances, this drag does not normally exceed the normal damping required for good servo stability.

Servo cylinders

Two important considerations in the selection of a servo cylinder is the leakage flow and the breakaway pressure which is in fact a measure of the pressure required to generate the necessary breakaway force. The rod is usually sealed with V-type and O-ring type seals since they provide reasonable resistance to external leakage.

Servo transducers

The function of a transducer is to convert a source of energy from one form to the other (for example from mechanical to electrical). In a servo system a feedback transducer after measuring the control system output generates a signal that is in turn fed back into the system for comparison with the input.

Transducers are also used in servo operations for instrumentation purposes, in order to measure the various parameters. Some of the important considerations in the selection of a transducer are the accuracy levels required, resolutional ability and repeatability. Transducers are generally categorized into digital and analog types. They may also be classified on the basis of their function as:

- Velocity transducers
- Pressure transducers
- Positional transducers
- Flow transducers and
- Acceleration transducers.

There are two basic types of servo valves that are widely used. They are:

1. Mechanical-type servo valve
2. Electro hydraulic servo valve.

6.4.1   Mechanical-type servo valve

Figure 6.41 shows a typical mechanical servo valve construction.

This valve is essentially a mechanical force amplifier used for positioning control. In this design, a small impact force shifts the spool by a specified amount. The fluid flows through port P_1, retracting the hydraulic cylinder to the right. The action of the feedback link shifts the sliding sleeve to the right until it blocks off the flow to the hydraulic cylinder. Thus, a given input motion produces a specific and controlled amount of output motion. Such a system where the output is fed back to modify the input, is called a closed loop system.

One of the most common applications of this type of mechanical hydraulic servo valve is in the hydraulic power steering system of automobiles and other transport vehicles.

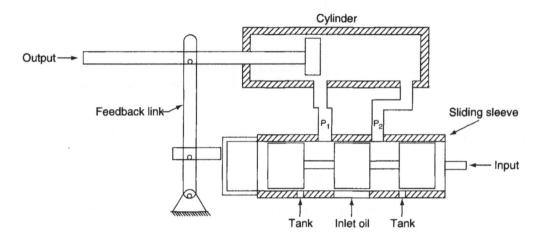

Figure 6.41
Mechanical hydraulic servo valve

6.4.2   Electro-hydraulic servo valve

In recent years, the electro-hydraulic servo valve has well and truly arrived on the industrial scene. The main characteristic of an electro-hydraulic valve is that its hydraulic output flow amplitude is directly proportional to the amplitude of its electrical DC input current. Typical electro-hydraulic valves use an electrical torque motor, a double nozzle pilot stage and a sliding spool second stage.

The torque motor includes components such as coils, pole pieces, magnets and an armature. The armature is supported for limited movement by a flexure tube. The flexure tube also provides a fluid seal between the hydraulic and electromagnetic portions of the valve. The flapper attaches to the center of the armature and extends down, inside the flexure tube. A nozzle is located at each side of the flapper such that the flapper motion varies the nozzle opening. Pressurized hydraulic fluid is supplied to each nozzle through an inlet orifice located at the end of the spool. A 40-micron screen that is wrapped around the shank of the spool, filters this pilot stage flow. The differential pressure between the ends of the spool is varied by the flapper motion between the nozzles.

The four-way valve spool directs flow from the supply to either control port C_1 or C_2 in an amount proportional to the spool displacement. The spool contains flow-metering slots in the control lands that are uncovered by the spool motion. Spool movement deflects a feedback wire that applies a torque to the armature/flapper. Electric current in the torque motor coil causes either clockwise or anti-clockwise torque on the armature. This torque displaces the flapper between the two nozzles. The differential nozzle flow moves the spool to either the right or left. The spool continues to move until the feedback torque counteracts the electromagnetic torque. At this point, the armature/flapper is returned to the center, the spool stops and remains displaced until the electrical input changes to a new level, thus making the valve spool position proportional to the electrical signal.

A simple description of the overall operation of an electro-hydraulic system can be made by referring to the following block diagram (Figure 6.42).

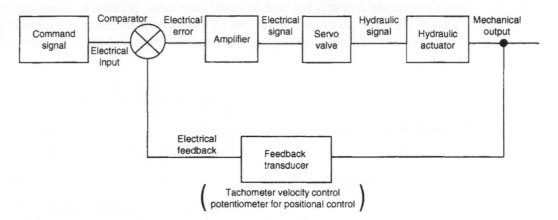

Figure 6.42
Block diagram of an electro-hydraulic servo system

The electro-hydraulic servo valve operates from an electrical signal to its torque motor, which positions the spool of a direction control valve. The signal to the torque motor comes from an electrical device such as a potentiometer. The signal from the potentiometer is electrically amplified to drive the torque motor of the servo valve. The hydraulic flow of the servo valve powers an actuator, which in turn drives the load. The velocity or position of the load is fed back in the form of an electrical input to the servo valve via a feedback device such as tachometer generator or potentiometer. Since the loop gets closed with this action, it is termed a closed loop system.

These servo valves are effectively used in a variety of mobile vehicles and industrial control applications such as earth moving vehicles, articulated arm devices, cargo handling cranes, lift trucks, logging equipments, farm machinery, steel mill controls, etc.

6.5   Hydraulic fuses

A hydraulic fuse is analogous to an electric fuse and its application in a hydraulic system is much the same as that of an electric fuse in an electrical circuit. A simple illustration of a hydraulic fuse is shown in Figure 6.43.

A hydraulic fuse when incorporated in a hydraulic system, prevents the hydraulic pressure from exceeding the allowable value in order to protect the circuit components from damage. When the hydraulic pressure exceeds the design value, the thin metal disk ruptures, to relieve the pressure and the fluid is drained back to the tank. After rupture, a new metal disk needs to be inserted before the start of the operation.

Hydraulic fuses are used mainly in pressure-compensated pumps with fail-safe overload protection, in case the compensator control on the pump fails to operate. A hydraulic fuse is analogous to an electric fuse because they both are 'one-shot' devices. On the other hand, the pressure relief valve is analogous to an electrical circuit breaker because they both are resetable devices.

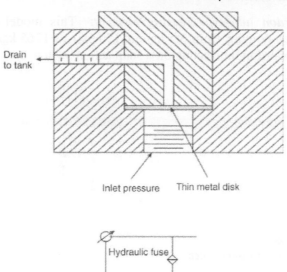

Drain
to tank

Inlet pressure   Thin metal disk

Hydraulic fuse

Figure 6.43
Hydraulic fuse

6.6 Pressure and temperature switches

6.6.1 Pressure switches

A pressure switch is an instrument that automatically senses a change in pressure and opens or closes an electrical switching element, when a predetermined pressure point is reached. A pressure-sensing element is that part of a pressure switch that moves due to the change in pressure. There are basically three types of sensing elements commonly used in pressure switches:

1. *Diaphragm*: This model (Figure 6.44) can operate from vacuum pressure up to a pressure of 10.5 kg/cm^2 (150 psi). It consists of a weld-sealed metal diaphragm acting directly on a snap action switch.

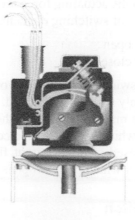

Figure 6.44
Diaphragm pressure switch

2. *Bourdon tube-type sensing element*: This model (Figure 6.45) can operate with pressures ranging from 3.5 kg/cm^2 (50 psi) to 1265 kg/cm^2 (18 000 psi). It has a weld-sealed bourdon tube acting on a snap action switch.

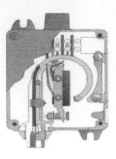

Figure 6.45
Bourdon tube pressure switch

3. *Sealed piston-type sensing element*:   This type of sensing element can operate with pressures ranging from 1 kg/cm^2 (15 psi) to 844 kg/cm^2 (12 000 psi). It consists of an O-ring-type-sealed piston direct acting on a snap action switch (Figure 6.46).

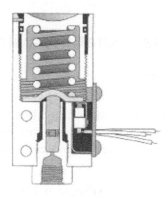

Figure 6.46
Sealed piston pressure switch

The electrical switching element in a pressure switch, opens and closes an electrical circuit in response to the actuating force received from the pressure-sensing element.

There are two types of switching elements:

1. Normally open
2. Normally closed.

A normally open switching element is one in which the current can flow through the switching element only when it is actuated. The plunger pin is held down by a snap action leaf spring and force must be applied to the plunger pin to close the circuit. This is done by an electrical coil which generates an electromagnetic field, when current flows through it. In a normally closed switch, current flows through the switching element until the element is actuated, at which point it opens and breaks the current flow.

6.6.2   Temperature switch

A temperature switch is an instrument that automatically senses a change in temperature and opens or closes an electrical switching element when a predetermined temperature level is reached. Figure 6.47 is an illustration of a common type of temperature switch which has an accuracy of ±1 °F maximum.

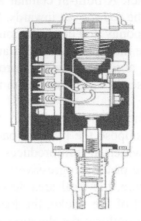

Figure 6.47
Temperature switch

This temperature switch is provided with an adjustment screw at the top end in order to change the actuation point. In order to facilitate its mounting on the hydraulic system whose temperature is to be measured, the bottom end of the switch is provided with threads. As in the case of pressure switches, temperature switches can also be wired either normally open or normally closed.

6.7 Shock absorbers

A shock absorber is a device, which brings a moving load to a gentle rest through the use of metered hydraulic fluid. Figure 6.48 shows the cut away section of a common type of shock absorber.

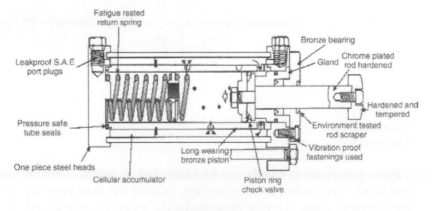

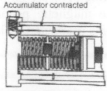

Figure 6.48
Shock absorber (Courtesy of EGD Inc.)

These shock absorbers are mounted, complete with oil. Therefore, they may be mounted in any position or angle. The spring return units are entirely self-contained units and extremely compact. A built-in cellular accumulator accommodates the oil displaced by the piston rod as the rod moves inwards. Since it is always filled with oil, there are no air pockets to cause spongy and erratic action.

Shock absorbers are multiple orifice hydraulic devices. When a moving load strikes the bumper of the shock absorber, it sets the rod and piston in motion. The moving piston pushes oil through a series of holes from an inner high-pressure chamber to an outer low-pressure chamber.

The resistance to the oil flow caused by the restrictions, creates a pressure, that acts against the piston to oppose the moving load. Holes are spaced geometrically according to a proven formula which in turn produces a constant pressure on the side of the piston opposite the load. The piston progressively shuts off these orifices as the piston and rod move inward. Therefore, the total area decreases continually while the load decelerates uniformly. At the end of the stroke, the load comes to a rest and the pressure drops to zero. This results in a uniform deceleration and gentle stopping with no bounce back. In bringing a moving load to a stop, the shock absorber converts work and kinetic energy into heat, which is dissipated to the surroundings.

One application of shock absorbers is in the energy dissipation of moving cranes. Here shock absorbers prevent bounce back of the bridge or trolley. The most common applications of shock absorbers are the suspension systems of automobiles.

6.8   Flowmeters

Flowmeters are used to measure flow in a hydraulic circuit. As shown in Figure 6.49, flowmeters mainly comprise of a metering cone and a magnetic piston along with a spring, for holding the magnetic piston in the no-flow position.

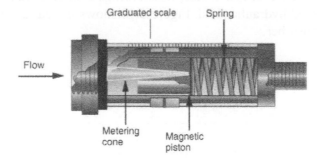

Figure 6.49
Flowmeter

Flow meters are normally not bi-directional in nature. They in fact act as check valves and block flow in the reverse direction. Initially the fluid entering the device flows around the metering cone, exerting pressure on the magnetic piston and spring. With increase in flow in the system, the magnetic piston begins compressing the spring and thereby indicates the flow rate on a graduated scale.

6.9   Manifolds

Leaky fittings are a cause for concern in hydraulic circuits especially with increase in the number of connections. This is where manifolds play a very important role. Their incorporation in a hydraulic circuit helps drastically reduce the number of external connections required. Figure 6.50 shows a simple manifold commonly employed in hydraulic systems.

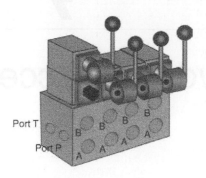

Figure 6.50
Manifold

In the case of modular valve stacking, the manifolds used are provided with common pressure and return ports, with each valve station being incorporated with individual A and B work ports. Manifolds are normally specified according to system pressure, total flow, number of work stations and valve size and pattern.

7

Hydraulic accessories

7.1   Objectives

After reading this chapter, one will be able to:

- Know the various accessories used in a hydraulic system
- Understand the function and construction of a reservoir
- Understand various types of accumulators from their design and construction point of view
- Select and specify accumulators for various applications
- Understand the concept of heat exchangers and their functions
- Know the specifications and construction of pipes and hoses
- Select and size hoses and pipes for different hydraulics applications.

7.2   Introduction

When we talk about hydraulics, it is not only fluids that come into one's mind. Also a discussion on hydraulics is not complete with only a discussion on pumps, motors and valves. There are other important components and aggregates in a hydraulic system listed under the category of hydraulic accessories. These accessories provide a clean and uninterrupted supply of fluid to a hydraulic system. In this chapter let us concentrate on learning how these accessories contribute not only to an efficient but also effective hydraulic system.

7.3   The reservoir system

The 'reservoir' as the name suggests, is a tank which provides uninterrupted supply of fluid to the system, by storing the required quantity of fluid. The hydraulic fluid is considered the most important component in a hydraulic system or in other words its very heart. Since the reservoir holds the hydraulic fluid, its design is considered quite critical. The reservoir in addition to storing the hydraulic fluid, performs various other important functions such as dissipating heat through its walls, conditioning of the fluid by helping settle the contaminants, aiding in the escape of air and providing mounting support for the pump and various other components. These are discussed in detail below. Some of the essential features of any good reservoir include components such as:

- Baffle plate for preventing the return fluid from entering the pump inlet
- Inspection cover for maintenance access

- Filter breather for air exchange
- Protected filler opening
- Level indicator for monitoring the fluid level
- Connections for suction, discharge and drain lines.

The proper design of a suitable reservoir for a hydraulic system is essential to the overall performance and life of the individual components. It also becomes the principle location where the fluid can be conditioned in order to enhance its suitability. Sludge, water and foreign matter such as metal chips have a tendency to settle in the stored fluid while the entrained air picked up by the oil is allowed to escape in the reservoir. This makes the construction and design of hydraulic reservoirs all the more crucial.

Many factors are taken into consideration when selecting the size and configuration of a hydraulic reservoir. The volume of the fluid in a tank varies according to the temperature and state of the actuators in the system. The volume of fluid in the reservoir is at a minimum with all cylinders extended and a maximum at high temperatures with all cylinders retracted. Normally a reservoir is designed to hold about three to four times the volume of the fluid taken by the system every minute. A substantial space above the fluid in the reservoir must be included to allow volume change, venting of any entrapped air and to prevent any froth on the surface from spilling out.

A properly designed reservoir can also help in dissipating the heat from the fluid. In order to obtain maximum cooling, the fluid is forced to follow the walls of the tank from the return line. This is normally accomplished by providing a baffle plate in the centerline.

The level of fluid in a reservoir is critical. If the level is too low, there is a possibility of air getting entrapped in the reservoir outlet pipeline going to the pump suction. This may lead to cavitation of the pump resulting in pump damage.

The monitoring of the temperature of the fluid in the reservoir is also important. At the very least, a simple visual thermometer whose ideal temperature range is around 45 °C (113 °F) to 50 °C (122 °F), needs to be provided on the reservoir.

There are basically two types of reservoirs:

1. Non-pressurized reservoir
2. Pressurized reservoir.

7.3.1   Non-pressurized reservoir

As the name suggests this type of reservoir is not pressurized, which means, the pressure in the reservoir will at no point of time rise above that of atmospheric pressure. Very extensively used in hydraulic systems, these reservoirs are provided with a vent to ensure that the pressure within, does not rise above the atmospheric value.

Figure 7.1 shows the typical construction of such a reservoir conforming to industry standards.

These reservoirs are constructed with welded steel plates. The inside surfaces are painted with a sealer, to prevent the formation of rust which might in turn occur due to the presence of condensed moisture. The bottom plate is sloping and contains a drain plug at its lowest point, to allow complete draining of the tank when required. In order to access all the internals for maintenance, removable covers are provided. A level indicator which is an important part of the reservoir, is also incorporated. This allows one to see the actual level of the fluid in the reservoir, while the system is in operation. A vented breather cap with an air filter screen helps in venting the entrapped air easily. The breather cap allows the tank to breathe when the fluid level undergoes changes in tune with the system demand.

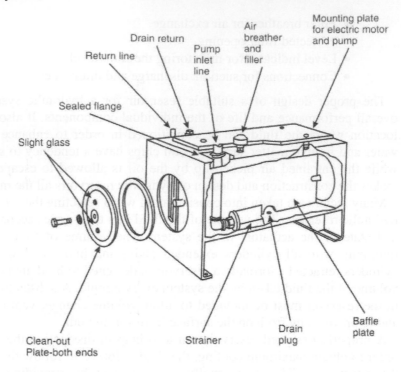

Figure 7.1
Non-pressurized reservoir

The baffle plate in the reservoir extends lengthwise across the center of the tank. Figure 7.2 shows a cross-sectional view of the reservoir depicting the baffle plate function.

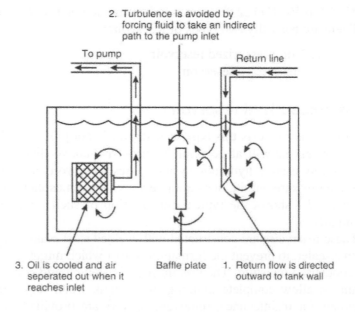

Figure 7.2
Baffle plate controls direction of flow in a non-pressurized reservoir

The height of the baffle plate in the reservoir is about 70% of the maximum fluid height. The purpose of the baffle plate is to separate the pump inlet line from the return line. This is done to prevent the same fluid from circulating continuously within the tank. In this way it is ensured that all the fluid is uniformly used by the system.

Essentially the baffle plate performs the following functions:

- It permits foreign substances to settle at the bottom
- It allows entrained air to escape from the fluid
- It prevents localized turbulence in the reservoir
- It promotes heat dissipation from the reservoir walls.

The reservoir is designed and constructed to facilitate the installation of a pump and motor on its top surface. A smooth machined surface of adequate strength is provided to support and maintain the alignment of the two units.

The return line enters the reservoir from the side of the baffle plate, which is opposite to the pump suction line. It should be below the fluid surface level all the time, in order to prevent foaming of the fluid. Similarly, the strainer or the foot valve should be located well below the normal fluid level in the reservoir and at least 1 in. or 2.5 cms above the bottom of the reservoir. If the strainer is located too high, it will lead to the formation of a vortex or crater that will permit ingress of air into the pump suction line.

The sizing of the reservoir is based on the following criteria:

- It should have sufficient volume and space to allow the dirt and metal chips to settle and the air to escape freely.
- It should be capable of holding all the fluid that might be drained from the system.
- It should be able to maintain the fluid level high enough to prevent air escaping into the pump suction line.
- The surface area of the reservoir should be large enough to dissipate the heat generated by the system.
- It should have sufficient free board over the fluid surface to allow thermal expansion of the fluid.

For most hydraulic systems, a reservoir having a capacity of three times the volumetric flow rate of the pump has been found to be adequate.

7.3.2 Pressurized reservoirs

Although it has been observed that non-pressurized reservoirs are the most suitable ones in a hydraulic system, certain hydraulic systems need to have pressurized reservoirs due to the nature of their application. For example, the Navy's aircraft and missile hydraulic systems essentially need pressurized reservoirs in order to provide a positive flow of fluid at higher altitudes where lower temperatures and pressure conditions are encountered.

Air-pressurized reservoir

The required pressure in the reservoir is maintained by means of compressed air. Compressed air is generally introduced into the reservoir from the top at a pressure specified by the manufacturer. In order to control this pressure, a pressure control device such as a pressure regulator is provided in the airline entering the reservoir. The function of this pressure regulator is to maintain a constant pressure in the reservoir, irrespective of the level and temperature of fluid in the reservoir.

A pressurized reservoir will only have a single entry point for filling up the fluid in the tank. Since the reservoir is always maintained at a pressure, it becomes important to have a foolproof system with safety relief valves, for the filling of fluid in the reservoir. Sufficient guidelines are provided by all manufacturers of such pressurized reservoirs.

7.4   Filters and strainers

7.4.1   Introduction

A modern hydraulic system must be highly reliable and provide greater levels of accuracy in its operation. The key to this is the requirement for high precision-machined components. Cleanliness of the hydraulic fluid is a vital factor in the efficient operation of the fluid power components. With the close tolerance design of pumps and valves, hydraulic systems are being made to operate at increased pressure and efficiency levels. The cleanliness of the fluid is an essential prerequisite for these components to perform as designed and also for higher system reliability and reduced maintenance.

The worst enemy of these high-precision components is contamination of the fluid. Essentially, contamination is the presence of any foreign material in the fluid, which results in detrimental operation of any of the components in a hydraulic system. Fluid contamination may be in the form of a liquid, gas or solid and can be caused by any of the following:

Built into the system during component maintenance and assembly

The contaminants here include metal chips, bits of pipe threads, tubing burrs, pipe dope, shreds of plastic tape, bits of seal material, welding beads, etc.

Generated within the system during operation

During the operation of a hydraulic system, many sources of contamination exist. They include moisture due to water condensation in the reservoir, entrained gases, scale caused by rust, bits of worn-out seal material, sludge and varnish due to oxidation of oil.

Introduced into the system from the external environment

The main source of contamination here is the use of dirty maintenance equipment such as funnels, rags and tools. Washing of disassembled components in dirty oil can also contaminate the fluid.

The foreign particles which are induced into the hydraulic system often get grounded into thousands of fine particles. These minute particles tend to lodge into the space between the control valve spools and their bores, causing the valve to stick. This phenomenon is called silting.

In order to keep the fluid free from all these contaminants and also in order to prevent phenomena such as silting, devices called filters and strainers are used in the hydraulic system. In this section, let us study in detail on how these filters keep the system clean and also dwelve on related topics such as micron rating, beta ratio and ISO code cleanliness levels.

7.4.2   Filters

A filter is a device whose primary function is to remove insoluble contaminants from the fluid, by use of a porous medium. Filter cartridges have replaceable elements made of nylon cloth, paper, wire cloth or fine mesh nylon cloth between layers of coarse wire. These materials remove unwanted particles, which collect on the entry side of the filter element. When saturated, the element is replaced. The particles sizes removed by the filters are measured in microns. One micron is one-millionth of a meter or 0.000039 of an inch. Filters can remove particles as small as 1 μ. Studies have proved that particle sizes as low as 1 μ can have a damaging effect on hydraulic systems and can also accelerate oil deterioration. Figures 7.3 gives details of the relative sizes of microscopic particles magnified 500 times.

Relative Sizes	
Lower limit of visibility (naked eye)	40 μ
White blood cells	25 μ
Red blood cells	8 μ
Bacteria	2 μ
Linear Equivalents	
1 in.	25 400 μ
1 mm	1000 μ

Mesh per Linear inch	US Sieve Number	Opening in inches	Opening in microns
52.36	50	0.0117	297
72.45	70	0.0083	210
101.01	100	0.0059	149
142.86	140	0.0041	105
200.00	200	0.0029	74
270.26	270	0.0021	53
323.00	325	0.0017	44

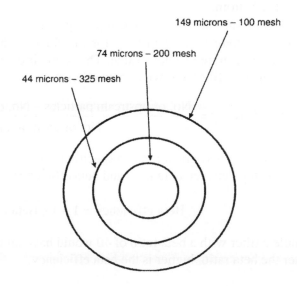

Figure 7.3
Relative mesh sizes

Micron (μ)

The particle sizes or clearances in hydraulic systems are usually designated in terms of micron which is equal to 39 millionths of an inch. To further simplify the process of understanding the concept of the micron, the smallest dot that can be seen by the naked eye is about 40 μm.

Beta ratio

It is a measure of a filter's efficiency. It is defined as the number of particles upstream from the filter that are larger than the micron rating of the filter, divided by the number of particles downstream from the filter larger than the micron rating of the filter. The example illustrated below demonstrates the concept of beta ratio quite clearly.

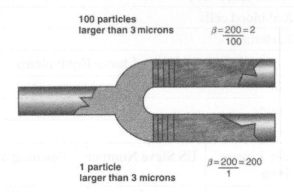

100 particles
larger than 3 microns $\beta = \dfrac{200}{100} = 2$

1 particle
larger than 3 microns $\beta = \dfrac{200}{1} = 200$

Figure 7.4
Comparison of filter efficiencies through beta ratio

From Figure 7.4, it is seen that there are 200 particles upstream from the filter which are larger than 3 μm in size. A filter having a lower beta ratio is less efficient because it allows more particles through it. Again referring the example above, it is seen that while the filter at the top allows 100 particles through, only 1 particle is allowed to pass through the filter at the bottom.

The beta ratio for the filter at the top is given by $\beta = 200/100 = 2$ which is a less efficient value, whereas the beta ratio for the filter at the bottom is given by $\beta = 200/1 = 200$ which is a more efficient value. The following equation is used to determine the efficiency value of a filter, known as beta efficiency.

$$\text{Beta efficiency} = \frac{\text{No. of upstream particles} - \text{No. of downstream particles}}{\text{No. of upstream particles}}$$

where the particle size is greater than a specified value of N μm.

The relationship between beta ratio and beta efficiency can thus be represented as:

$$\text{Beta efficiency} = 1 - 1\sqrt{\text{Beta ratio}}$$

For example a filter with a beta ratio of 40 would have an efficiency of 1/40 = 97.5%. The higher the beta ratio, higher is the beta efficiency.

Fluid cleanliness level

Fluid cleanliness can be defined according to ISO, NAS and SAE standards. ISO 4406 defines contamination levels using a dual scale numbering system. The first number refers to the quantity of particles over 5 μ per 100 ml of fluid and the second number refers to the number of particle over 15 μ per 100 ml of oil.

For example, a cleanliness level of 15/12 indicates that there are between 2^{14} and 2^{15} particles over 5 μ and 2^{11} and 2^{12} particles over 15 μ, per 100 ml of fluid.

Type of System	Minimum, Recommended Cleanliness Level			Minimum Recommended Filtration Level (μ)
	ISO 4406	NAS 1638	SAE 749	
Silt sensitive	13/10	4	1	2
Servo	14/11	5	2	3–5
High pressure (250–400 bar)	15/12	6	3	5–10
Normal pressure (150–250 bar)	16/13	7	4	10–12
Medium pressure (50–150 bar)	18/15	9	6	12–15
Low pressure (<50 bar)	19/16	10	–	15–25
Large clearance	21/18	12	–	25–40

Let us consider another example based on a 1 ml fluid sample. A particle count analysis is done for this sample using specific particle sizes of 4 μm, 6 μm and 14 μm. By selecting these three sizes, an accurate assessment of the amount of silt from 4 μm and 6 μm particles can be obtained, while the number of particles above 14 μm is an indication of the amount of wear type particles in the fluid.

No. of particles per 1.0 ml

ISO 4406 code

Scale No.	More than	Up to
0	–	–
1	–	–
5	–	–
7	–	–
13	40	80
15	–	–
18	1.3 k	2.5 k
20	–	–
22	20 k	40 k

From the ISO 4406 table shown, let us consider a rating of 22/18/13. This indicates the following:

1. 22 indicates that the number of particles greater than or equal to 4 μm in size is more than 20 000 and less than or equal to 40 000, per ml.
2. 18 indicates that the number of particles greater than or equal to 6 μm in size is more than 1300 and less than or equal to 2500, per ml.
3. 13 indicates that the number of particles greater than or equal to 14 μm in size is more than 40 and less than or equal to 80, per ml.

The ISO contamination code is applicable to all fluid types and provides a universal expression of relative cleanliness, between the suppliers and the users of hydraulic fluid. This code is meaningful, only if it is related to the required cleanliness level of the hydraulic system under consideration. It is usually based on the manufacturer's requirements for the cleanliness levels in which a component may operate. For example, most gear pumps may sufficiently operate in fluids having a rating of 18/16/15 ISO, while servo valves require an ISO code 15/13/12 or better.

Filter location

Filter location in a hydraulic system is critical to ensuring acceptable levels of fluid cleanliness and adequate component protection. Figure 7.5 shows the location of the various filters in a hydraulic system.

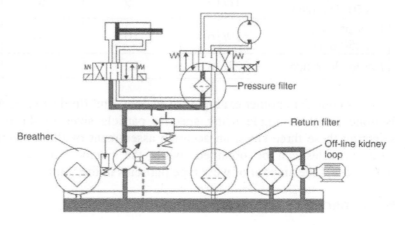

Figure 7.5
Location of filters in a hydraulic circuit

The function of breathers in a hydraulic system is to prevent entry of airborne particles which are drawn into the system due to changes in the fluid level of the reservoir. They are usually mounted on the reservoir. Components such as servo valves which are located immediately downstream of the filter are protected from wear and silting-related problems by pressure filters. These pressure filters are designed to withstand high pump pulsations and the system pressure. Return line filters provide protection against entry of particulate matter when the fluid is returned to the tank. An off-line filter also known as kidney loop is often provided in a hydraulic system especially when fluid circulation through the return-line filter is minimal. Off-line filters operate on a continuous basis. The chief advantage associated with these filters is the flexibility they offer with regard to their placement. Since these filters are independent of the main system, their location in a hydraulic circuit can be chosen in such a way that easy serviceability is ensured.

7.4.3   Filtration methods

There are three basic types of filtering methods used in hydraulic systems.

Mechanical

This type normally contains a metal, a cloth screen, or a series of metal disks separated by thin spacers. Mechanical type filters are capable of removing only relatively coarse particles from the fluid.

Absorbent type

These filters are porous and contain permeable materials such as paper, wood pulp, diatomaceous earth, cloth, cellulose and asbestos. Paper filters are normally impregnated with a resin, to provide added strength. In this type of filter, the particles are actually absorbed as the fluid permeates the material. As a result this method is used for extremely small particle filtration.

Adsorbent type

Adsorption is a surface phenomenon and refers to the tendency of the particles to cling to the surface of the filter. Thus the capacity of such a filter depends on the amount of surface area available. Adsorbent materials used, include activated clay and chemically treated paper.

7.4.4   Types of filters

Some filters are designed to be installed in the pressure line and are normally used in systems where high-pressure components such as valves are more dirt sensitive than the pump. Return line filters are used in systems, which do not have large reservoirs to permit contaminants to settle at the bottom. A return line filter is needed in systems containing close tolerance high-performance pumps. Let us now study some of the common filters normally used in hydraulic systems.

Duplex-type filters

Figure 7.6 shows a filter designed for either a suction line or a pressure line. This is a duplex filter.

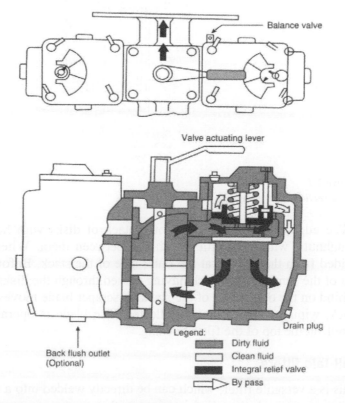

Figure 7.6
Cross-section of a duplex filter

A duplex filter, as the name suggests consists of two filters out of which only one is in use all the time. When the filter element gets clogged, the second filter is put to use. Dirty fluid comes into the middle section and passes down through the filter element. The filter element can be of fine gage nylon cloth or wire cloth or finely perforated stainless steel.

From the filter element, the fluid passes out of the unit and into the line. This unit has a 'telltale' indicator, which indicates when the filter element is excessively clogged and requires cleaning. If the filter is not cleaned after indication, the fluid bypasses the filter element and there is no filtering action. Such a bypass mechanism is important for a filter because, when the filter element is clogged heavily, the pump in line may get damaged due to starvation of hydraulic fluid.

Edge-type filters

This is also referred to as a full flow filter, which means that all the oil in the system passes through it. Figure 7.7 illustrates an edge-type filter.

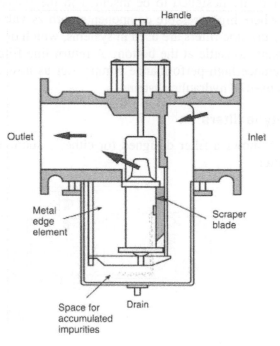

Figure 7.7
Cross-section of an edge-type filter

The edge-type filter consists of a stack of disks with holes in their center, like flat doughnuts, with very little separation between them. When entering the filter, fluid is guided from the bottom at the outer side of the stack. Before leaving the filter, it comes out of the center of the stack having passed through the disks. Impurities in the oil are left behind on the outer edge of the stack. A scraper blade moves over the outer surface of the stack, wiping off all the dirt collected. The blade is operated manually by means of a handle at the top of the filter housing.

Tell-tale filters

This is a versatile filter, which can be directly welded into a reservoir's suction and return line or conversely installed in pipes with a maximum pressure of 10 kg/cm^2 (142 psi). This filter can remove particles as small as 3 μ. It consists of an indicating element, which indicates the time when cleaning is required. That is why this filter is referred to as a tell-tale filter (Figure 7.8).

The operation of a tell-tale filter is dependent on the fluid passing through the porous media, which traps the contaminants. The tell-tale indicator monitors the pressure differential buildup due to dirt, which gives an indication of the condition of the filter element.

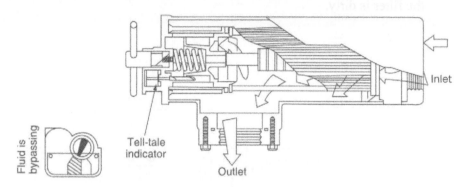

Figure 7.8
Cross-section of a tell-tale filter (Courtesy of Parker Hannifin Corp., Michigan)

Principle of working

Fluid enters the inlet at the bottom of the tube, passes from the inside to the outside through the filter element and exits at the side outlet. When a new and recently cleaned filter element is fitted in the housing, the tell-tale indicator will indicate that it is 'clean'.

As dirt deposits on the surface of the element, the pressure differential across the inlet and outlet of the element rises. The bypass piston senses this difference in pressure. The piston is held seated by a spring. When the element requires cleaning, the pressure differential is high enough to compress the spring, forcing the piston off its seat. The piston movement causes the tell-tale indicator to point towards the 'needs cleaning' position.

When the filter element is not cleaned when this signal comes on, the pressure differential continues to rise, causing the piston to uncover a bypass passage in the cover. This action limits the rise in the pressure differential to a value equal to the spring tension and the fluid bypasses the filter element. A cutaway view of a tell-tale filter is shown in Figure 7.9.

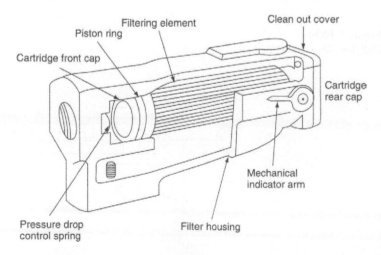

Figure 7.9
Cutaway view of a tell-tale filter

The following schematics show a few of the typical methods used for filtration. Figure 7.10(a) shows the location of a proportional flow filter. As the name implies, proportional flow filters are exposed to only a percentage of the total flow in the system. The primary disadvantage of this type of filtration arrangement is that, there is no positive protection of any specific component within the system and there is no way to know that the filter is dirty.

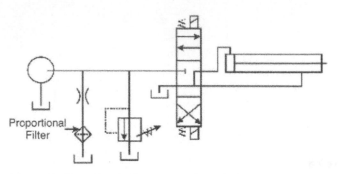

Figure 7.10(a)
Proportional flow filter in a separate drain line

Figures 7.10(b)–(d) show full flow filtration in which all the flow from the pump is accepted.

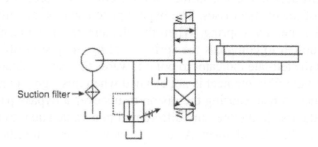

Figure 7.10(b)
Full flow filter in suction line

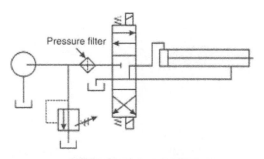

Full flow filter in pressure line

Figure 7.10(c)
Full flow filter in pressure line

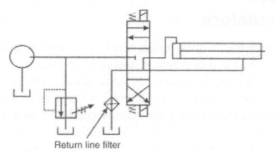

Full flow filter in return line

Figure 7.10(d)
Full flow filter in return line

7.4.5 Strainers

A strainer is a device made of wire mesh screens, which seek to remove large solid particles from a fluid. As part of standard engineering practice, strainers are installed on pipelines ahead of valves, pumps and regulators, in order to protect them from the damaging effects of fluid and other system contaminants.

A common strainer design uses two screens, cylindrical in shape. One cylinder is inside the other and the two are separated by a small space. The outer cylinder is a coarse mesh screen and the inner one is a fine mesh screen. The fluid first passes through the coarse mesh screen and filters the larger particles. It then passes through the fine mesh screen, which blocks the smaller particles. Figure 7.11 shows the cross-sectional view of a typical strainer.

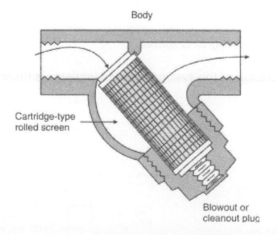

Figure 7.11
Strainer

The bottom of the strainer serves as the sump (or pot) for the solids to collect. The strainer can be cleaned out easily at intervals, by two different procedures:

1. The cleanout plug can be removed and the pressure in the line can be used to blow the fixture clean.
2. The large retaining nut at the bottom is to be removed for pulling the mesh out of the strainer in order to clean it and putting it back in line.

7.5   **Accumulators**

Accumulators are devices, which simply store energy in the form of fluid under pressure. This energy is in the form of potential energy of an incompressible fluid, held under pressure by an external source against some dynamic force. This dynamic force can come from three different sources: gravity, mechanical springs or compressed gases. The stored potential energy in the accumulator is the quick secondary source of fluid power capable of doing work as required by the system. This ability of the accumulators to store excess energy and release it when required, makes them useful tools for improving hydraulic efficiency, whenever needed. To understand this better, let us consider the following example.

A system operates intermittently at a pressure ranging between 150 bar (2175 psi) and 200 bar (2900 psi), and needing a flow rate of 100 lpm for 10 s at a frequency of one every minute. With a simple system consisting of a pump, pressure regulator and loading valves, this requires a 200 bar (2900 psi), 100-lpm pump driven by a 50 hp (37 kW) motor, which spends around 85% of its time, unloading to the tank. When an accumulator is installed in the system as shown in Figure 7.12, it can store and release a quantity of fluid at the required system pressure.

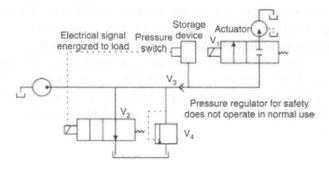

Figure 7.12
Circuit diagram showing an accumulator

The operation of the system with accumulator is illustrated by Figure 7.13;

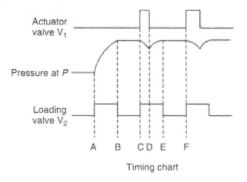

Figure 7.13
Graphical representation of accumulator operation

At time A, the system is turned on and the pump loads, causing pressure to rise as the fluid is delivered to the accumulator via a non-return valve V_3. At time B, the working pressure is reached and a pressure switch on the accumulator causes the pump to unload. This state is maintained as the non-return valve holds the system pressure.

The actuator operates between time C and D. This draws the fluid from the accumulator causing a fall in the system pressure. The pressure switch on the accumulator puts the pump on load again, to recharge the accumulator for the next cycle.

With the accumulator in the system, the pump now only needs to provide 170 lpm and also requires reduced motor hp. Thus it can be seen how an accumulator helps in reducing the power requirements in a hydraulic system.

There are three basic types of accumulators used extensively in hydraulic systems. They are:

1. Weight-loaded or gravity-type accumulator
2. Spring-loaded-type accumulator
3. Gas-loaded-type accumulator.

7.5.1   Weight-loaded-type accumulators

The weight-loaded type is historically the oldest type of accumulator. It consists of a vertical heavy wall steel cylinder, which incorporates a piston with packing to prevent leakage (Figure 7.14).

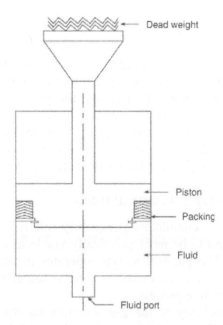

Figure 7.14
Weight-loaded accumulator

A dead weight is attached to the top of the piston. The gravitational force of the dead weight provides the potential energy to the accumulator. This type of accumulator creates a constant fluid pressure throughout the full volume output of the unit, irrespective of the rate and quantity. The main disadvantage of this accumulator is its extremely large size and heavy weight.

7.5.2   Spring-loaded-type accumulators

A spring-loaded accumulator is similar to the weight-loaded type except that the piston is preloaded with a spring. A typical cross-section of this type of accumulator has been illustrated in Figure 7.15.

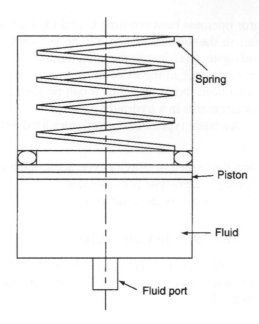

Figure 7.15
Spring-loaded accumulator

The spring is a source of energy, acting against the piston and forcing the fluid into the hydraulic system. The pressure generated by this accumulator depends on the size and preloading of the spring. In addition, the pressure exerted on the fluid is not constant. They typically deliver small volumes of oil at low pressures and therefore tend to be heavy and large for high-pressure, large volume systems.

A spring-loaded accumulator should not be used for applications requiring high cycle rates as the spring may lose its elasticity and render the accumulator useless.

7.5.3   Gas-loaded-type accumulators

These types of accumulators (frequently referred to as hydro-pneumatic accumulators) have been found to be more practically viable as compared with the weight and spring-loaded types. The gas-loaded type operates in accordance with Boyle's law of gases, according to which the pressure of a gas is found to vary inversely with its volume for a constant temperature process.

The compressibility of the gas accounts for the storage of potential energy in these accumulators. This energy forces the oil out of the accumulator when the gas expands, due to a reduction in system pressure.

Gas-loaded accumulators fall under two main categories:

    1. Non-separator type
    2. Separator type.

7.5.4   Non-separator type

The non-separator type consists of a fully enclosed shell containing an oil port at the bottom and the gas-charging valve at the top. The valve is confined to the top and the oil to the bottom of the shell. There is no physical separator between the gas and oil, and thus the gas pushes directly on the oil.

The main advantage of this type of accumulator is its ability to handle a large volume of oil. However, its disadvantage lies in the fact that the oil tends to absorb gas due to the lack of a separator. A cross-section of a non-separator type accumulator has been illustrated in Figure 7.16.

A gas-loaded accumulator must be installed vertically to keep the gas confined to the top of the accumulator. It is not recommended for use with high-speed pumps as the entrapped gas in the oil may cause cavitation and damage the pump. The absorption of gas in the oil also makes the oil compressible, resulting in spongy operation of the actuators.

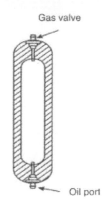

Figure 7.16
Non-separator-type gas loaded accumulator

7.5.5   Separator type

This is the most commonly accepted design under gas-loaded accumulators. In this type there is a physical barrier between the gas and the oil. This barrier effectively utilizes the compressibility property of the gas.

The separator type accumulator is in turn classified into three types:

1. The piston type
2. The diaphragm type and
3. The bladder type.

Piston-type separator gas-loaded accumulator

This accumulator consists of a cylinder containing a freely floating piston with proper seals, as illustrated in Figure 7.17.

The piston serves as a barrier between the gas and oil. A threaded lock ring provides a safety feature that prevents the operator from disassembling the unit while it is precharged.

The main disadvantage of piston-type accumulators is that they are very expensive and have size limitations. In low-pressure systems, the piston and seal friction also poses problems. Piston accumulators should not be used as pressure pulsation dampeners or shock absorbers because of the inertia of the piston and the friction in the seals.

The principle advantage of the piston-type accumulator lies in its ability to handle very high- or low-temperature system fluids, through the utilization of compatible O-ring seals.

Diaphragm-type separator gas-loaded accumulator

The diaphragm-type accumulator consists of a diaphragm secured in a shell and serving as an elastic barrier between the oil and the gas. The cross-sectional view of a diaphragm-type accumulator is shown in Figure 7.18.

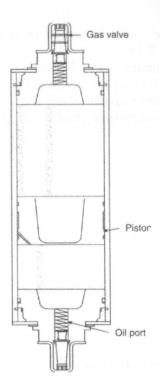

Figure 7.17
Piston-type accumulator

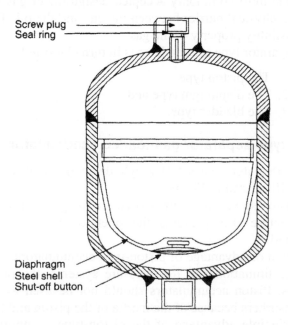

Figure 7.18
Diaphragm-type accumulator

A shut off button which is secured at the base of the diaphragm, covers the inlet of the line connection when the diaphragm is fully stretched. This prevents the diaphragm from being pressed into the opening during the precharge period. On the gas side, the screw plug allows control of the charge pressure and the charging of the accumulator by means of a charging and testing device.

With the help of the following figures (Figures 7.19(a)–(f)), let us now see how exactly a diaphragm-type accumulator works.

Figure 7.19(a) shows the accumulator without the nitrogen charge in it or in other words in a precharged condition. The diaphragm can be seen in a non-pressurized condition.

Figure 7.19(b) shows the accumulator in charged condition. Here nitrogen is charged into the accumulator, to the precharged pressure.

Figure. 7.19(c) shows how the hydraulic pump delivers oil to the accumulator and how this process leads to the deformation of the diaphragm.

As seen from Figure 7.19(d), when the fluid delivered reaches the maximum required pressure, the gas is compressed. This leads to a decrease in gas volume and subsequent storage of hydraulic energy.

Figure 7.19(e) shows the discharge of the oil back to the system when the system pressure drops, indicating requirement of oil to build back the system pressure.

Figure 7.19(f) shows the accumulator attaining its original precharged pressure condition.

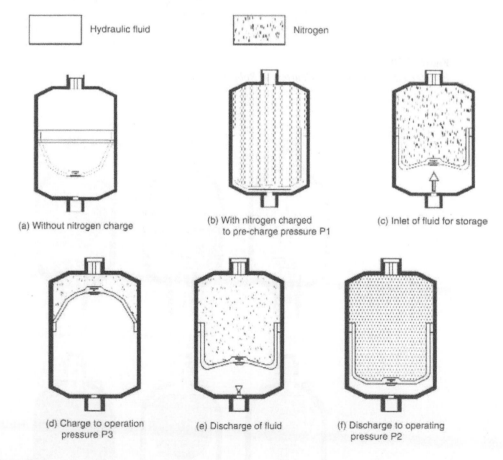

Figure 7.19
Operation of a diaphragm-type accumulator

The primary advantage of the diaphragm-type accumulator is the small weight-to-volume ratio, which makes it highly suitable for airborne applications.

Bladder-type separator gas-loaded accumulator

The bladder-type accumulator contains an elastic barrier between the oil and gas as shown in the cross-sectional view in (Figure 7.20).

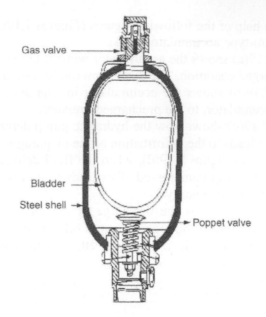

Figure 7.20
Bladder-type accumulator (Courtesy of Robert Bosch Corp.)

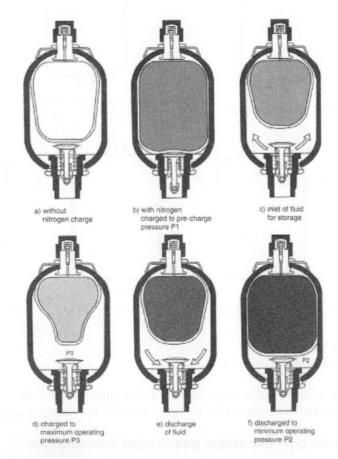

a) without nitrogen charge

b) with nitrogen charged to pre-charge pressure P1

c) inlet of fluid for storage

d) charged to maximum operating pressure P3

e) discharge of fluid

f) discharged to minimum operating pressure P2

Figure 7.21
Operation of a bladder-type accumulator

The bladder is fitted to the accumulator by means of a vulcanized gas-valve element that can be installed or removed through the shell opening at the poppet valve. The poppet valve closes the inlet when the bladder is fully expanded. This prevents the bladder from being pressed into the opening. A shock-absorbing device, protects the valve against accidental shocks, during a quick opening.

The greatest advantage with these accumulators is the positive sealing between the gas and oil chambers. The lightweight bladder provides a quick pressure response for pressure regulation as well as applications involving pump pulsations and shock dampening.

Figure 7.21 illustrates the functioning of a bladder-type accumulator.

The hydraulic pump delivers oil to the accumulator and deforms the bladder. As the pressure increases, the volume of gas decreases. This results in the storing of hydraulic energy. Whenever additional oil is required by the system, it is supplied by the accumulator even as the pressure in the system drops by a corresponding amount.

7.5.6 Accumulator applications

From a study of the above, we have understood the principle of operation and functioning of various types of accumulators used in hydraulic systems. Let us now discuss them from the point of view of their application. Accumulators are mainly used:

- As auxiliary power sources
- As leakage compensators and
- As hydraulic shock absorbers.

An auxiliary power source

This is one of the most common applications of an accumulator. In this application, the purpose of the accumulator is to store the oil delivered by the pump during the work cycle. The accumulator then releases the stored oil on demand, to complete the cycle, thereby serving as a secondary power source to assist the pump. In such a system where intermittent operations are performed, the use of an accumulator results in reduced pump capacity.

Figure 7.22 outlines this application with the help of symbols.

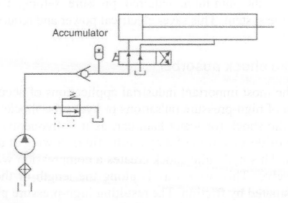

Accumulator

Figure 7.22

Accumulator as an auxiliary power source

In this application, a four-way valve is used in conjunction with an accumulator. When the four-way valve is manually actuated, oil flows from the accumulator to the blank end of the cylinder. This extends the piston until the end of the stroke. When the cylinder is in a fully extended position, the pump charges the accumulator. The four-way valve is then de-activated to retract the cylinder. Oil flows from the pump and the accumulator to retract the cylinder rapidly. This is how an accumulator can be used as an auxiliary power source.

Leakage compensator

In this application (Figure 7.23), the accumulator acts as a compensator, by compensating for losses due to internal or external leakage that might occur during an extended period of time, when the system is pressurized, but not in operation.

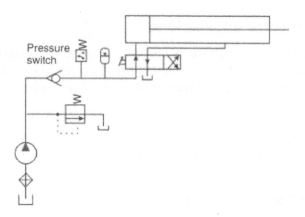

Figure 7.23
Accumulator as a leakage compensator

The pump charges the accumulator and the system, until the maximum pressure setting on the pressure switch is obtained. When the system is not operating, it is required to maintain the required pressure setting, to accomplish which the accumulator supplies leakage oil to the system during a lengthy period of time. Finally when the system pressure falls below the minimum required pressure setting, the pump starts to automatically recharge the system. This saves electrical power and reduces heat in the system.

Hydraulic shock absorber

One of the most important industrial applications of accumulators is in the elimination or reduction of high-pressure pulsations or hydraulic shocks.

Hydraulic shock (or water hammer, as it is frequently called) is caused by the sudden stoppage or deceleration of a hydraulic fluid flowing at a relatively higher velocity in the pipelines. This hydraulic shock creates a compression wave at the location of the rapidly closing valve. This wave travels along the length of the entire pipe, until its energy is fully dissipated by friction. The resulting high-pressure pulsations or high-pressure surges may end up damaging the hydraulic components.

An accumulator installed near the rapidly closing valve as shown in Figure 7.24 can act as a surge suppressor to reduce these high-pressure pulsations or surges.

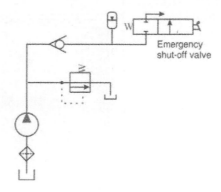

Figure 7.24
Accumulator as a hydraulic shock absorber

7.6 Heat exchangers

Heat is generated in a hydraulic system because of the simple reason that no component can operate at 100% efficiency. Significant sources of heat include pumps, pressure relief valves and flow control valves. This can cause a rise in temperature of the hydraulic fluid above the normal operating range. Heat is continuously generated whenever the fluid flows from a high-pressure region to a low-pressure region, without producing mechanical work. Excessive temperatures hasten oxidation of the hydraulic fluid and also reduce its viscosity. This promotes deterioration of seals and packings and accelerates wear and tear of hydraulic components such as valves, pumps and actuators. This is the reason why temperature control is a must in hydraulic systems.

The steady-state temperature of the fluid depends on the rate of heat generation and the rate of heat dissipation. If the fluid-operating temperature is excessive, it means that the rate of heat dissipation is inadequate for the system. Assuming that the system is reasonably efficient, the solution is to increase the rate of heat dissipation. This is accomplished by the use of 'coolers', which are commonly known as heat exchangers. In certain applications, the fluid needs to be heated in order to achieve the required viscosity of the fluid in the system. For example, if a mobile hydraulic equipment is required to operate in sub-zero conditions, the fluid needs to be heated. In such cases, heat exchangers are termed as heaters.

The factors to be considered when sizing a heat exchanger are:

- The required drop in temperature of the hydraulic fluid
- The flow of the hydraulic fluid in the system
- The time required to cool the fluid.

There are two main types of heat dissipation heat exchangers:

1. Air-cooled heat exchangers and
2. Water-cooled heat exchangers.

7.6.1 Air-cooled heat exchanger

Figure 7.25 shows an air-cooled heat exchanger.

The hydraulic fluid to be cooled is pumped through the tubes that are finned. As the fluid flows through the tubes, air is blown over them. This takes away the heat from the tubes. A fan driven by an electric motor is incorporated in the heat exchanger to provide air for cooling. The heat exchanger shown above, uses tubes which contain special devices called turbulators whose function is to mix the warmer and cooler oils for better heat transfer.

Figure 7.25
Air-cooled heat exchanger

Advantages associated with air-cooled heat exchangers are:

1. Substantial cost reduction because of the use of air for cooling purposes, as compared with water
2. Lower installed costs
3. Possibility of the dissipated heat being reclaimed.

Disadvantages of air-cooled heat exchangers are:

1. Relatively larger in size
2. High noise levels
3. Higher installation costs.

7.6.2   Water-cooled heat exchanger

Figure 7.26 is an illustration of a common type of water-cooled heat exchanger used in hydraulic systems.

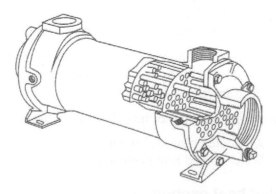

Figure 7.26
Water-cooled heat exchanger

This is typically a shell and tube-type heat exchanger. The cooling water is pumped into the heat exchanger and flows around the tube bank. The hydraulic fluid, which is to be cooled, flows through the tubes. While flowing through the tubes, the fluid gives away heat to the water, thereby reducing its temperature.

Advantages of water-cooled heat exchangers are:

1. They are very compact and cost-effective
2. They do not make noise
3. They are good in dirty environments.

Disadvantages associated with water-cooled heat exchangers are:

1. Water costs can be expensive
2. Possibility of mixing of oil and water in the event of rupture
3. Necessity for regular maintenance to clear mineral deposits.

7.7 Fluid conductors – hydraulic pipes and hoses

7.7.1 Introduction

Efficient transmission of power from one location to the other is a key element in the design and performance of a hydraulic system. This is known as fluid conducting. Fluid conductors comprise that part of the hydraulic system that is used to carry fluid to the various components. These conductors include the likes of steel tubing, steel pipes and hydraulic hoses.

In an actual hydraulic system, the fluid flows through a distribution network consisting of pipes and fittings, which carry the fluid from the reservoir through the operating components and back to the reservoir. Since power is transmitted throughout the system by means of this network, it is highly imperative that this network be properly designed to ensure efficient operation.

7.7.2 Steel piping

Steel pipes are often preferred over other conductors from the standard point of performance and cost. However since welding is required to be carried out on these pipings to ensure maximum leak protection, they are quite difficult to assemble. Another factor to be reckoned with is that they require costly flushing during start-up, to ensure a contaminant free environment. Although steel pipes are specified by their nominal outside diameter, their actual flow capacity is determined by their inside area.

One can classify a piping system into two types:

1. Metallic piping
2. Non-metallic piping.

The choice regarding the type of piping to be used for a system, primarily depends on its operating pressures and flow rates. Additionally, it also depends on environmental conditions such as the type of fluid, operating temperatures and atmospheric conditions.

Let us study the various types of pipes and their fittings used in a hydraulic system.

7.7.3 Metal piping

A pipe is a device used to convey fluids from one point to another without any physical movement on its part. Metallic pipes are the ones made from iron or steel. In order to have a common platform for specifying pipes all over the world, pipe sizes are standardized and are usually expressed in inches or millimeters. As a rule the size of a pipe is given in terms of its outside diameter or inside diameter.

Figure 7.27 shows the cut section of a pipe and the terminology associated with it.

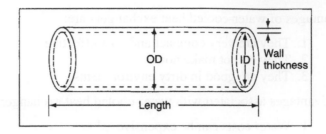

Figure 7.27
Pipe size terminology

The principal dimensions with regard to pipes are:

- Outside diameter
- Inside diameter
- Wall thickness and
- Length.

Commercial pipelines up to 12 in. (305 mm) in size are designated by their inside diameters while pipe sizes above 12 in. (305 mm) are designated by their outside diameters.

Different metals have different characteristics, which render them usable for a variety of applications. An alloy is a metal made up of two or more metals, which dissolve into each other when melted together. Mixing a metal with a non-metal can also form an alloy.

Metals are generally referred to as ferrous or non-ferrous metals. A ferrous metal is one that contains iron, while a non-ferrous metal does not contain iron. Piping is commonly made of wrought iron, cast iron or steel. The difference between the three is largely the carbon content in each of them. In addition to the common ferrous and non-ferrous metals used in piping material, there are some exclusive metals that are used in special piping applications. Aluminum pipes are light in weight and corrosion resistant, although they experience a decrease in strength with increase in temperature. Lead pipes are also considered suitable for highly corrosive fluids. Piping made from special materials is more expensive. Stainless steel, a common type of ferrous metal, is an alloy of steel and chromium.

7.7.4 Pipe schedules and codes

In the past, all piping was designated as standard, extra strong and double extra strong. That system allowed no variation for the wall thickness. Also, as the piping requirements started increasing, a greater variation was needed for specifying the pipes. As a result, piping today is classified according to a schedule. The most common schedule numbers in practice are 40, 80, 120 and 160. For pipe diameters ranging from 1/8 in. (15 mm) to 10 in. (250 mm), the dimensions of standard steel pipe correspond to schedule 40.

The standard steel pipes correspond to schedule 40 pipes with a wall thickness of 6 mm. The dimensions of extra strong steel pipes are the same as schedule 80 pipes ranging from diameters 15 to 200 mm.

For schedule numbers ranging from 10 to 160, the basic difference is the wall thickness. When the wall thickness of any given pipe size is increased, the inside diameter of the pipe decreases.

Due to the increasing variety and complexity of requirements for piping, a number of engineering societies and standards have devised codes, standards and specifications that cater to a vast majority of applications.

Some codes provide the formulae for determining the minimum pipe size and the minimum wall thickness for use in a given application. Some codes provide information on the pipe material, method of manufacture (seamless or welded), ASTM (American Society of Testing and Materials) specification number, the grade of pipe and the stress that can be exerted on the pipe at various conditions of pressure and temperature. By consulting these codes, a designer can determine exactly the type of specification for a particular application. Selecting a pipe according to code recommendations is similar to the procedure adopted by an automobile mechanic who refers the operation and maintenance manual to determine the type of oil filter that is to be fitted for a particular car.

The pipe schedules commonly used in hydraulic systems are 40, 80 and 160. The classification has been done based on their nominal size and schedule numbers.

Nominal Size in inches	Pipe Outside Diameter in inches (mm)	Pipe Inside Diameter		
		Schedule 40	Schedule 80	Schedule 160
1/8	0.405 (10.3)	0.269	0.215	
1/4	0.540 (13.7)	0.364	0.302	
3/8	0.675 (17.1)	0.493	0.423	
1/2	0.840 (21.3)	0.622	0.546	0.466
3/4	1.050 (26.6)	0.824	0.742	0.614
1	1.315 (33.4)	1.049	0.957	0.815
1–1/4	1.660 (42.1)	1.380	1.278	1.160
1–1/2	1.900 (48.2)	1.610	1.500	1.338
2	2.375 (60.3)	2.067	1.939	1.689

The ANSI (American National Standards Institute) has established a code for the identification of pipelines. This code involves the use of legends, nameplates or tags and colors. The code states that, 'Identification of the contents of a piping system shall be by a lettered legend giving the name of the contents in full or in abbreviated form'. All the pipes are color coded with respect to their content, for easy identification. As shown in the table below, the use of colors provides a general indication of the type of material carried by the pipe.

Color identification code

Classification	Color Field	Color of Letters for Legend
Fire quenching material	Red	White
Inherently hazardous material	Yellow	Black
Inherently low hazardous material		
Liquid	Green	White
Gas	Blue	White

7.7.5   Pipe fittings

The components used in a piping system to connect the various sections of the pipe in order to change the direction of flow, are called fittings. Fittings are made from a number of materials including steel, bronze, cast iron, plastic and glass.

Standards have been established to ensure that fittings are made from proper materials and are able to handle the designed pressures. Some of the important functions of these fittings are:

- Changing the direction of flow
- Providing branch connections
- Changing the size of lines
- Closing lines
- Connecting lines.

Fittings for changing the direction of flow

To change the direction of flow, the fittings that are normally used are referred to as elbows. Elbows generally come in all angles but those that are commonly used are the 45° and 90° ones.

Figure 7.28 shows two common types of elbows used in industrial hydraulic applications.

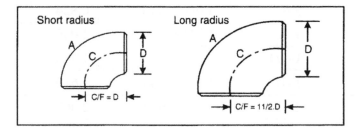

Figure 7.28
Short and long radius elbows

A large radius fitting has the more gradual curve of the two. Along with the diameter and schedule of the pipe, elbows are also specified with center-to-face dimensions. It is the distance between the center of the fitting (A) and a line (C) drawn down from the face (D) of the fitting at the other end. In the elbow of larger radius, the center-to-face distance is always 1.5 times the diameter of the fitting. This type of elbow is used in applications where the rate of flow is critical and there is limitation regarding space. In a short radius elbow, the center-to-face distance is equal to the pipe diameter.

Fittings for changing the size of the pipelines

The function of the reducer is to reduce the line to a smaller pipe size. One reason for doing so is to increase the pressure in the system.

As in the case of elbows, we can see from Figure 7.29 that there are different types of fittings that can be used in a piping system, in order to meet the above requirements.

7.7.6   Pipe joints

Pipe joints can be of the screwed, flanged or welded type. Each of these joints which are widely used, have their own advantages and disadvantages.

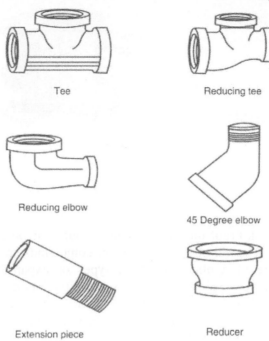

Figure 7.29
Various fittings used for a piping system

Screwed fittings are joined to the pipe by means of threads. The main advantage of the threaded joints is that the pipe length can be easily tailored at a later stage. As a threaded joint has metal-to-metal contact between the threads of two mating parts of the pipe, the risk of leakage is high. Hence sealing of the threaded joint becomes very important. The traditional method of providing a seal between the mating threads is to coat the threads with a paste dope. In recent years, the invention of Teflon sealing tapes has made for a more effective solution. The Teflon (being a registered trademark of Du Pont De Neumors and Company) tape can be simply wound over the threads for sealing.

Flanged fittings are either forged or made as cast pipe fittings. A flange is a rim or ring at the end of the fitting, which mates with another section of the pipe. Pipe sections can also be made with flanged ends. Flanges are joined either by being bolted together or by being welded together. The flanged faces again have metal-to-metal contact and proper sealing needs to be provided between the two mating surfaces to avoid leakages. A gasket is usually inserted between the mating surfaces of two flanges, which are bolted together. Normally, compressed asbestos gaskets can be used under conditions of normal pressure and temperature. However, when the system operates under higher pressure and temperature conditions, higher-grade gaskets are used.

Other pipe fittings used in flanged connections include expansion joints and vibration dampeners. Expansion joints have three functions:

1. They compensate for slight changes in the length of the pipes, by allowing the joined sections of rigid pipes to expand or contract with changes in temperature.
2. They allow pipe motion either to the side or along the length of the pipe, as the pipe shifts after installation.
3. They help dampen vibration to some extent and reduce the noise carried along the pipe and originating from distant pumps.

A typical expansion joint is shown in Figure 7.30.

Figure 7.30
Expansion joint

It has a leak proof tube that extends through the bore and forms the outside surface of the flanges. Natural or synthetic rubber compounds are used for this purpose, depending on the type of application. Other types of expansion joints include metal bellows (corrugated type), spiral wound types and slip joint types.

Vibration dampeners are designed specifically to absorb vibration because vibrations reduce the life of the pipes as well as the operating equipments. They also eliminate line noises carried by the pipes.

A typical vibration damper is shown in Figure 7.31.

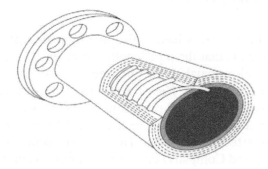

Figure 7.31
Vibration dampener

7.8   Steel tubing

Steel tubing is used in hydraulic systems when rigid lines are required. They are easier to assemble and don't require welding in order to achieve leak-proof connections. Seamless steel tubing is the most widely used conductor type for hydraulic systems as it has significant advantages over pipes. The tubing can be bent into any shape thereby reducing the number of fittings in a system. Tubing is easier to handle and can be reused without any sealing problems. For low-volume systems, tubing can handle the pressure and flow requirements with less bulk and weight. However the flip side to tubing and their fittings is that that they are very expensive.

Steel tubes are measured and specified by their outside diameter and wall thickness. The tubing grade and wall specification determine the pressure ratings. As shown in Figure 7.32, one piece of tubing is connected to another component through a tube connector and fastening nut. The tube is often pre-flared to an angle of 37° to accept a 37° flare connector as shown in Figure 7.32.

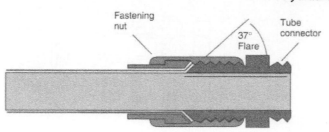

Figure 7.32
Steel tubing arrangement

Some of the most common tube sizes used in hydraulic systems have been tabulated below.

Tube OD inches	Wall Thickness in mm	Tube ID in mm	Tube OD	Wall Thickness in mm	Tube ID in mm
1/8	0.89	1.5	5/8	0.89	14.0
3/16	0.89	3.0	3/4	1.3	16.5
1/4	0.89	4.5	7/8	1.3	19.8
5/16	0.89	6.2	1	1.65	22.9
3/8	0.89	7.8	1–1/4	1.65	28.5
1/2	0.89	10.9	1–1/2	1.65	34.8

The most widely used material for steel tubing is SAE1010 dead soft, cold drawn steel. This material has a fairly high tensile strength and is quite easy to work with. In order to obtain higher tensile strengths, the tubes are made of AISI 4130 steel.

The fittings required for tubing are slightly different to those required for piping. Although the concept of fittings remains the same, the method of joining and sealing are different. A few fittings along with the sealing methods adopted for tubing have been illustrated in Figure 7.33.

Some of these fittings are known as compression fittings. They seal by metal-to-metal contact and may be either a flared type or flare-less type. Other fittings may use O-rings for sealing purposes.

The sleeve inside the nut, supports the tube, to dampen vibrations. When the hydraulic component has straight threaded ports, straight thread type O-ring fittings can be used. This type of sealing is ideal for higher pressures, because as the pressure increases the seal gets tighter.

The tubing is supported ahead of the ferrules, by the fitting body. Two ferrules grasp tightly around the tube without causing any damage to the tube wall. There is virtually no constriction of the inner wall, ensuring minimum flow restriction.

Figure 7.34 shows a Swage tube fitting which can sustain any pressure up to the bursting strength of the tubing, without leakage. This type of fitting can be repeatedly taken apart and re-assembled.

The secret of the Swagelok fitting is that all the fitting action is along the axial direction of the tube and not rotary. Since no torque is transmitted from the fitting to the tubing, there is no formation of initial strain that may weaken the tubing.

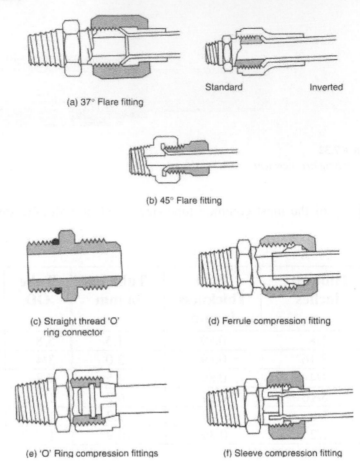

(a) 37° Flare fitting

Standard   Inverted

(b) 45° Flare fitting

(c) Straight thread 'O' ring connector

(d) Ferrule compression fitting

(e) 'O' Ring compression fittings

(f) Sleeve compression fitting

Figure 7.33
Threaded fittings and connectors used for tubing

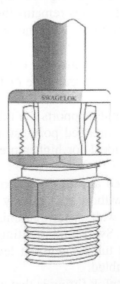

Figure 7.34
Swagelok tube fitting

The tensile strength of the fitting is that it retains a value virtually up to the internal strength of the line, since these connections/fittings can be repeatedly disassembled and re-assembled.

The screw-on type Swagelok fitting is that as the tube is swaged, the axial direction of the tube and not rotary. Since no torque is transmitted to the tube, the tubing there is no formation of initial strain that may weaken the tubing.

7.9   Flexible hoses

Flexible hoses are one of the most important conductors used in hydraulic systems. They are used in applications where lines must flex or bend or in other words when hydraulic components such as actuators are subjected to movement. Hoses are fabricated in layers of elastomer (synthetic rubber) and braided fabric or braided wire, which permit operation at high pressure.

Normally, hoses are rated by a safety factor between 1and 4. The various types of coatings and reinforcements used, determine the specific pressure ratings. The volume and velocity of the fluid flow determines the hose size and unlike pipes and tubings, hoses are designated by the inside diameter.

The construction of a typical flexible hose is illustrated in Figure 7.35.

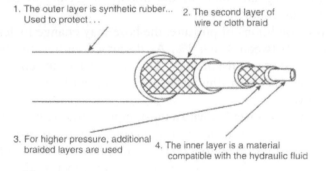

Figure 7.35
Construction of a flexible hose

The outer layer is normally made of synthetic rubber and serves to protect the braid layer. The hose can have as few as three layers (one being the braid). When multiple layers are used, they may alternate with synthetic rubber layers or the wire layers may be directly placed over one another. Hose construction has been standardized by the Society of Automotive Engineers under SAE J5-17 also known as the R series (an example being 100 R4) by which the cover, construction, application and pressure rating is described.

The table below gives typical hose sizes and dimensions for a single wire braided and double wire braided designs.

Hose Size	Tube OD inches	Single Wire Braided			Double Wire Braided		
		Hose ID inches	Hose OD inches	Minimum Bend Radius (inches)	Hose ID inches	Hose OD inches	Minimum Bend Radius (inches)
4	1/4	3/16	33/64	1–15/16	1/4	11/16	4
6	3/8	5/16	43/64	2–3/4	3/8	27/32	5
8	1/2	13/32	49/64	4–5/8	1/2	31/32	7
12	3/4	5/8	1–5/64	6–9/16	3/4	1–1/4	9–1/2
16	1	7/8	1–15/64	7–3/8	1	1–9/16	11
20	1–1/4	1–1/8	1–1/2	9	1–1/4	2	16

Size specifications of single wire braided hoses represent the outside diameter in one-sixteenths of an inch of standard tubing, with the hose having the same inner diameter as that of the tubing. For example, a size 8 single wire-braided hose will have an inside diameter very close to 8/16 or 1/2 in. standard tubing.

Care should be taken to change the fluid content in hoses since the hose and fluid material should be compatible. When installing the hoses, some amount of slackness is always to be ensured. This helps relieve stress. Although hoses, can last a long time, they are not as permanent as metal conductors because the rubber tends to deteriorate over a period of time due to contact with various substances such as solvents, sunlight, heat and water.

7.9.1   Hose routing and installation

The following are some of the important factors to be considered when installing a flexible hose.

Under conditions of pressure, the hose may change its length. This range in turn can be anywhere between 4 and 2%. As discussed earlier, the hose should always have some slackness during installation. This is needed to compensate for any shrinkage or expansion. However, this slackness should not be in excess as that might result in a bad installation.

If the hose is installed with a twist, high operating pressures tend to force it to straighten out. This can loosen the hose connectors and can also burst the hose at the point of maximum strain. Therefore this condition needs to be avoided.

When the hose line passes over an exhaust manifold, or any other heat source, it must be insulated by a heat resistant boot or a metal baffle. At locations having bends, the hose length should be sufficiently long for it to form a wide radius curve. Too small a bend can pinch the hose and restrict the flow. This condition also needs to be avoided.

8

Hydraulic fluids

8.1   Objectives

After reading this chapter, the student will be able to:

- Understand and explain the primary functions of a hydraulic fluid
- Detail the various characteristics of hydraulics fluids
- List the different types and categories of hydraulic fluids used
- Understand the problems encountered with hydraulic fluids
- Compare the common properties of the different hydraulic fluid types.

8.2   Introduction

The working fluid is the single most important component of any hydraulic system. It serves as a lubricant, heat transfer medium, Sealant and most important of all, a means of energy transfer. These fluid functions can be clearly understood with the help of the diagrams illustrated below. Fluid characteristics play a critical part in determining the equipment performance and life. Hydraulic fluids are basically non-compressible in nature and can therefore take the shape of any container. This tendency of the fluid makes it exhibit a certain advantage in the transmission of force across a hydraulic system. Use of a clean, high-quality fluid, is an essential prerequisite for achieving efficient operation of the hydraulic system. Although early hydraulic systems employed the medium of water for transferring hydraulic energy, there are serious limitations attached to it such as:

- Its relatively high freezing point (water freezes at 0 °C or 32 °F when the pressure is atmospheric)
- Its tendency to expand when frozen
- Its corrosive nature
- Its poor lubrication properties
- Its capacity to dissolve more oxygen leading to phenomena such as oxygen pitting.

This has necessitated the development of modern fluids designed specifically for application in hydraulic systems, a detailed discussion of which will follow.

Although hydraulic fluid types vary according to application, the four common types are:

1. Petroleum-based fluids which are the most common of all fluid types and widely used in applications where fire resistance is not required.
2. Water glycol fluids used in applications which require fire resistance fluids.
3. Synthetic fluids used in applications where fire resistance and non-conductivity is required.
4. Environment-friendly fluids that end up causing minimal effect on the environment in the event of a spill.

As discussed earlier, hydraulic fluids have the four essential primary functions of power transmission, heat dissipation, lubrication and sealing to accomplish which, they should possess the following properties:

1. Ideal viscosity
2. Good lubricity
3. Low volatility
4. Non-toxicity
5. Low density
6. Environmental and chemical stability
7. High degree of incompressibility
8. Fire resistance
9. Good heat-transfer capability
10. Foam resistance and most importantly
11. Easy availability and cost-effectiveness.

It is quite obvious that no single fluid can meet all the above requirements and it is therefore essential that only the fluid that comes closest to satisfying most of these requirements be selected for a particular application.

8.3   Characteristics of hydraulic fluids

In the first chapter of this book, we have examined in detail, the various properties of hydraulic fluids that help determine the system performance and efficiency. There are two other important characteristics, which also play an important role in the life of a hydraulic fluid. These are:

1. Oxidation and corrosion prevention
2. Neutralization number.

8.3.1   Oxidation and corrosion prevention

Oxidation is that process resulting from the chemical reaction of oxygen in the air with oil. This can reduce the life of a hydraulic fluid drastically. Petroleum oils are particularly susceptible to oxidation because oxygen readily unites with both carbon and hydrogen molecules. Most products of oxidation are soluble in oil as well as acidic in nature and can cause severe damage to system components by way of corrosion. The products of oxygen include insoluble gums, sludge and varnish and these tend to increase the viscosity of oil.

There are a number of parameters, which hasten the rate of oxidation once it begins, some of the important ones being heat, pressure, contaminants, water and metal surfaces.

However oxidation is most affected by temperature. Various additives are incorporated in hydraulic oils to inhibit the rate of oxidation. As additives increase the cost of the oil, they should be specified only if necessary, based on the temperature and other environmental conditions.

The relative changes in oil property under oxidizing conditions can be studied with the help of a standard recommended test. This test which is detailed below along with a graphic illustration, gives a measure of the formation of harmful products in oils (Figure 8.1).

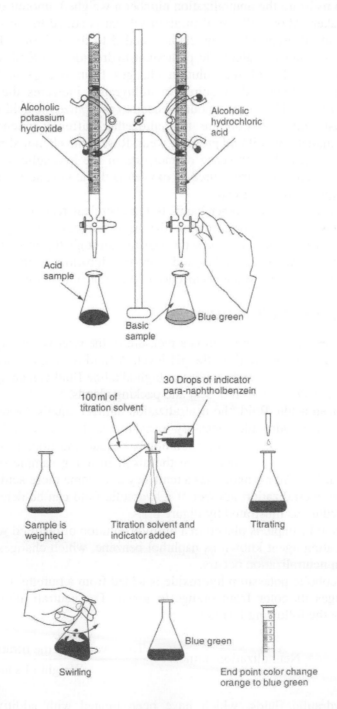

Figure 8.1
Oil oxidation test

The main purpose behind this test is to measure the resistance to oxidation by measuring the change in the acidity of the oil due to absorbed oxygen. The test procedure is as follows:

A 300 ml sample of oil is placed in a tube and immersed in an oil bath at 95 °C. Three liters per hour of oxygen is allowed to pass continuously through the sample for a period of about 1000 h. The acidity of oil is then measured by determining the neutralization number.

To measure the neutralization number a weighed amount of a sample of oil is placed in a beaker. About 100 ml of titration solvent is added to the sample in the beaker. Further around 30 drops of an indicator is added to this solution. For carrying out the titration process, standard alcoholic potassium hydroxide is added to the solution drop by drop, until the color of the solution changes from orange to blue-green. The amount of potassium hydroxide required in milligrams, indicates the level of oxidation that has taken place. This is the amount needed to neutralize the acid in one gram of oil.

Rust and corrosion are two altogether different phenomena, although they both contaminate the oil and promote wear. Rust is the chemical reaction between iron or steel and oxygen. The presence of moisture in the hydraulic system provides the necessary oxygen. One primary source of oxygen is the atmospheric air, which enters the reservoir through the breather cap.

Corrosion on the other hand, is the chemical reaction between a metal and an acid. Because of rusting or corrosion, the metal surfaces of the hydraulic components are eaten away. This results in excessive leakage through the affected parts like seals. Rust and corrosion can be resisted by additives, which form a protective layer on the metal surfaces and thereby prevent the occurrence of a chemical reaction.

8.4   Neutralization number

The neutralization number is a measure of the relative acidity or alkalinity of a hydraulic fluid and is specified by the pH level. A fluid having a smaller neutralization number is recommended, as high-acidity or high-alkaline fluid can cause corrosion of metal parts as well as a deterioration of seal and packing glands.

For an acidic fluid, the neutralization number equals the number of milligrams (mg) of potassium hydroxide necessary to neutralize the acid in a 1 g sample. In the case of an alkaline fluid, the neutralization number equals the amount of alcoholic hydrochloric acid that is necessary to neutralize the alkali in a 1 g sample of hydraulic fluid. With use, hydraulic fluid normally has a tendency to become more acidic than basic.

The neutralization number of a hydraulic fluid can be determined by the following test procedure as illustrated by Figure 8.2.

The oil sample is placed in a titration solution of distilled water, alcohol, toluene and an indicating agent known as naphthol benzene, which changes color from orange to green when neutralization occurs.

Alcoholic potassium hydroxide is added from a burette drop by drop, until the solution changes its color from orange to green. The neutralization number is then calculated using the following formula,

$$\text{Neutralization number} = \frac{\text{Total weight of the titrating solution} \times 5.61}{\text{Weight of sample used}}$$

Hydraulic fluids, which have been treated with additives in order to inhibit the formation of acids, are usually able to keep this number at a low value of 0 or 0.1.

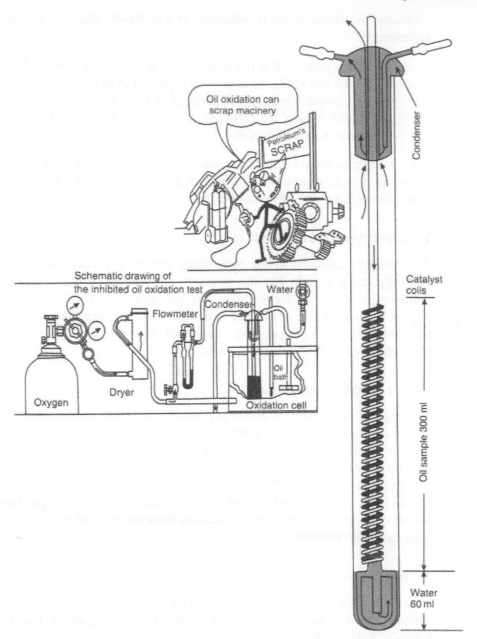

Figure 8.2
Neutralization number test

8.5 Fire-resistant fluids

It is important for a hydraulic fluid to neither initiate nor support fire. Most hydraulic fluids will however burn under certain conditions. There are many hazardous applications where concern for human safety demands the use of fire resistant fluids. Examples include coal mines, hot metal processing equipments, aircraft and marine fluid power systems.

A fire resistant fluid is one, which can be ignited but will not support a flame when the ignition source is removed. Flammability of a fluid is defined as the ease of ignition and the ability to propagate a flame.

In order to determine the flammability of a hydraulic fluid, the following characteristics are tested:

- *Flash point*:   It is the temperature at which the fluid surface gives off vapors, which can ignite when a flame is passed over it.
- *Fire point*:   It is the temperature of a fluid at which the fluid surface gives off vapors, which are sufficient to support combustion for a time of 5 s, when a flame is passed over it.

Various fire-resistant fluids have been developed in recent times, to reduce fire hazards. There are basically three types of fire-resistant fluids that are commonly used in hydraulic applications. They are discussed below.

8.5.1   Water–glycol solutions

This solution contains about 40% water and 60% glycol. These solutions have high viscosity index values, but the viscosity rises as the water evaporates. The operating temperature range of these fluids lies between −23 °C (−9.4 °F) and 83 °C (180 °F approx.). Most of the newer synthetic seal materials are compatible with water–glycol solutions. Metals such as zinc, cadmium and magnesium react with water–glycol solutions and hence should not be used.

8.5.2   Water-in-oil emulsions

This type of fluid contains about 40% water completely dispersed in a special oil base. It is characterized by small droplets of water completely surrounded by oil. Although water provides good coolant properties, it makes the fluid more corrosive. As a result a greater amount of corrosion inhibitor additives are necessary. The operating temperature range of this fluid lies between −28 °C (−18.4 °F) and 83 °C (180 °F approx.). Even in the case of this fluid, it is necessary to replenish the water to maintain proper viscosity of the fluid. These types of fluids are compatible with most rubber seal materials found in petroleum base hydraulic systems.

8.5.3   Straight synthetics

It is a chemically formulated fluid designed to inhibit combustion and generally has the highest fire-resistant temperature. Typical fluids belonging to this type are phosphate esters and chlorinated hydrocarbons.

The disadvantages of these types of fluids are their low viscosity index, incompatibility with most natural and synthetic rubber seals and high costs. In particular, the phosphate esters readily dissolve pipe thread compounds, paints and electrical insulation.

8.6   Foam-resistant fluids

Air can be dissolved or entrained in hydraulic fluids. For example, if the return line to the reservoir is not submerged, the jet of oil entering the liquid surface will carry air with it. This causes air bubbles to form in oil. If these bubbles rise to the surface too slowly, they can be drawn into the pump intake, leading to cavitation and subsequent pump damage.

Similarly, a small leak in the suction line can cause entrainment of large quantities of air from the atmosphere. This type of leak is difficult to detect since in this case air leaks in, rather than the oil leaking out from the suction line. Another adverse effect of entrained or dissolved air is a significant reduction in bulk modulus of the hydraulic fluid. This can have serious consequences in terms of stiffness and accuracy of hydraulic actuators. The amount of dissolved air can be significantly reduced by properly designing the reservoir, since this is where most of the air is picked up.

Another method is to use a premium grade hydraulic fluid that contains foam-resistant additives. These additives are chemical compounds, which break out entrained air and in the process quickly separate the air from the oil, in the reservoir itself.

8.7 General types of fluids

8.7.1 Petroleum-based fluids

The first major category of hydraulic fluids is the petroleum-based fluid, which is the most widely used type. The crude oil that is quality refined can be used for light services.

Additives should be added to these fluids in order to maintain the following characteristics:

- Good lubricity
- High viscosity index
- Oxidation resistance.

The primary disadvantage of a petroleum-based fluid is that it is flammable. In order to take care of this, fire-resistant hydraulic fluids have been developed, as already discussed in the beginning.

8.7.2 Lubricating oils

These are conventional engine type oils. Due to their better lubricating properties, they enhance the life of the hydraulic components. These oils contain anti-wear additives used to prevent engine wear on cams and valves. Their improved lubricity also provides wear resistance to heavily loaded hydraulic components such as pumps and valves.

8.7.3 Air

Air is also one of the fluids used in hydraulic systems. However, systems that use air as the medium are known as pneumatic systems. The advantages of using air are:

- Air does not burn.
- It can be easily made available in a clean form by the use of filters.
- Any leakage of air from the system is not messy as it simply breaks into the atmosphere.
- Air can also be made into an excellent lubricator by adding a fine mist of oil using a lubricator.
- Use of air in the system eliminates the return lines as air can be simply exhausted back to the atmosphere.

Air also has certain major disadvantages, some of them being:

- Its compressibility
- Its sluggishness and lack of rigidity
- Its corrosivity on account of the presence of oxygen and water.

To summarize, the single most important component in a fluid power system is the working fluid. No single fluid contains all the ideal characteristics required. The designer should select the fluid having the properties closest to that required by a particular application.

The properties of some of the common hydraulic fluids are tabulated below:

Fluid	Specific Gravity	Weight Density	Absolute Viscosity (lb/ft³)	Kinematic Viscosity (cst)
Hydraulic liquids				
Mineral oil	0.89	55.6	133.4	150.0
Water oil emulsion	0.90	56.2	149.0	166.0
Water glycol solution	1.10	68.6	110.0	100.0
Phosphate ester	1.10	68.6	220.0	200.0
Silicone oil	1.04	64.8	41.6	40.0
MIL 5606	0.86	53.6	19.1	22.0
Miscellaneous liquids				
Castor oil	0.97	60.5	986.0	1016.0
Ethyl alcohol	0.79	49.4	1.20	1.51
Ethylene glycol	1.12	69.9	19.9	17.8
Gasoline	0.68	42.5	2.64	3.88
Glycerol	1.26	78.6	1490.0	1180.0
Linseed oil	0.94	58.8	65.0	68.9
Mercury	13.6	849.0	1.55	0.114
Mineral oil SAE	0.91	56.7	114.0	125.3
Olive oil	0.92	57.1	84.0	91.8
Turpentine	0.87	54.3	1.49	1.71
Water	1.00	62.4	1.00	1.00

9

Applications of hydraulic systems

9.1 Objectives

After reading this chapter the student will be able to:

- Understand the arrangement of various components in a hydraulic system
- Understand the subject of hydraulics as applied to the following:

  - Hydraulic-powered and controlled sky tram
  - Bendix hydro boost brake system
  - Power steering
  - Welding
  - Bridge maintenance.

9.2 Introduction

There are essentially three ways of transmitting power:

1. Electrical
2. Mechanical
3. Fluid power.

Most applications actually use a combination of all these three means, to obtain an efficient overall system. In order to exactly determine which of the above methods is best suited to a particular application, it is important to know the salient features of each method. For example, hydraulic systems can transmit power more economically than mechanical systems, over a larger distance. As in the case with mechanical systems, hydraulic systems are not hindered by the geometry of components in the system.

Industry today is becoming increasingly dependent on automation, in order to increase productivity. Hydraulic or fluid power can be considered to be the 'muscle' of automation and is therefore being widely used in various applications. In the discussion to follow, we shall discuss the relative advantages of hydraulic systems and their various applications.

9.3   Advantages of hydraulic systems

A hydraulic system has four major advantages, which makes it quite efficient in transmitting power.

1. *Ease and accuracy of control*:   By the use of simple levers and push buttons, the operator of a hydraulic system can easily start, stop, speed up and slow down.
2. *Multiplication of force*:   A fluid power system (without using cumbersome gears, pulleys and levers) can multiply forces simply and efficiently from a fraction of a pound, to several hundred tons of output.
3. *Constant force and torque*:   Only fluid power systems are capable of providing a constant torque or force regardless of speed changes.
4. *Simple, safe and economical*:   In general, hydraulic systems use fewer moving parts in comparison with mechanical and electrical systems. Thus they become simpler and easier to maintain.

In spite of possessing all these highly desirable features, hydraulic systems also have certain drawbacks, some of which are:

- Handling of hydraulic oils which can be quite messy. It is also very difficult to completely eliminate leakage in a hydraulic system.
- Hydraulic lines can burst causing serious human injuries.
- Most hydraulic fluids have a tendency to catch fire in the event of leakage, especially in hot regions.

It therefore becomes important for each application to be studied thoroughly, before selecting a hydraulic system for it. Let us now discuss some of the most important and common hydraulic system applications.

9.4   Components of hydraulic systems

Virtually, all-hydraulic circuits are essentially the same regardless of the application. There are six basic components required for setting up a hydraulic system:

1. A reservoir to hold the liquid (usually hydraulic oil)
2. A pump to force the liquid through the system
3. An electric motor or other power source to drive the pump
4. Valves to control the liquid direction, pressure and flow rate
5. An actuator to convert the energy of the liquid into mechanical force or torque, to do useful work. Actuators can either be cylinders which provide linear motion or motors which provide rotary motion and
6. Piping to convey the liquid from one location to another.

Figure 9.1 illustrates the essential features of a basic hydraulic system with a linear hydraulic actuator.

The extent of sophistication and complexity of hydraulic systems vary depending on the specific application.

Each unit is a complete packaged power system containing its own electric motor, pump, shaft coupling, reservoir and miscellaneous piping, pressure gages, valves and other components required for operation. We have already reviewed the functions of all these components in the previous chapters.

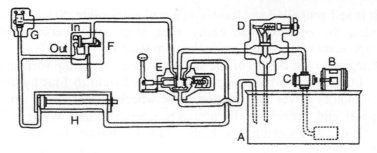

List of components

A – Reservoir
B – Electrical motor
C – Pump
D – Maximum pressure
    (relief) valve

E – Directional valve
F – Flow control valve
G – Right-angle check valve
H – Cylinder

Figure 9.1
Basic hydraulic system with a linear hydraulic actuator

9.5   Applications of hydraulic systems

The widespread use of fluid power in a vast majority of modern day applications is a testimony to its efficiency. Now that we are quite familiar with the design, functional and operational aspects of individual components in a hydraulic system, let us proceed further and discuss some of these common but important applications.

9.5.1   High wire hydraulically driven overhead tram

Most overhead trams require haulage or tow cable to travel up and down steep inclines. A 22-passenger, 12 000 pound (around 5000 kg) hydraulically powered and controlled tram is shown in Figure 9.2.

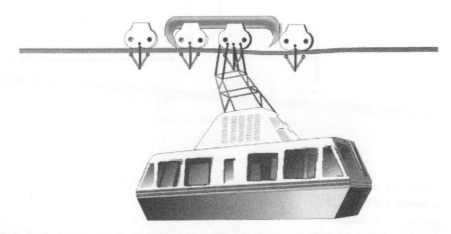

Figure 9.2
Hydraulic powered and controlled sky tram

It is self-propelled and travels on a stationary cable. Since the tram moves instead of the cables, the operator can easily start, stop and reverse a particular car completely independent of any other car in the tram system.

Integral to the design of the sky tram drive is a pump (driven by a standard 8 cylinder gasoline engine) which supplies pressurized fluid to four hydraulic motors. Each of the four motors drives two friction drive wheels. Eight drive wheels on top of the cables support and propel the tramcar. On steep inclines, while a higher driving torque is required for ascending, a higher braking torque is required during descent. Dual compensation of the four hydraulic motors provides efficient proportioning of the available horsepower to meet the variable torque demands.

9.5.2 Bendix hydro-boost brake system

This system was developed by Bendix Corporation as a solution to the typically crowded engine compartments consisting of larger vacuum units. Figure 9.3 contains a schematic of this system.

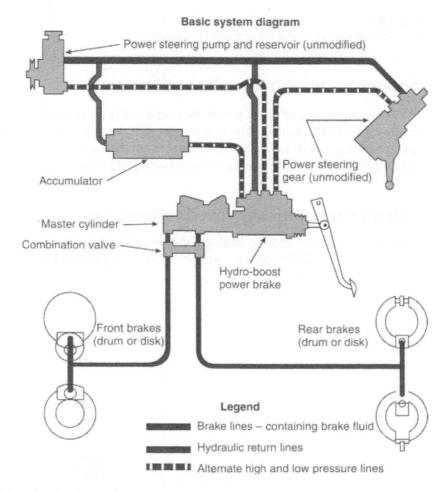

Figure 9.3
Bendix hydro-boost brake system

The basic system consists of an open center spool valve and a hydraulic cylinder assembled in a single unit. The power steering pump supplies the operating pressure. Hydro-boost provides power assist to operate the dual master cylinder braking system.

Normally mounted on the engine compartment firewall, it is designed to provide specific brake feel characteristics throughout the wide range of pedal force and travel. A spring accumulator stores energy for reverse stops.

9.5.3 Power steering

Power steering is another automotive application developed by Bendix Corporation. This is used in conjunction with a conventional type steering gear. The hydraulic power cylinder is mounted at any convenient place where it can be connected to act directly on the steering tie rod or equivalent linkage member (Figure 9.4). Power for steering is applied in the most simplest and direct way as straight-line motion to the existing steering linkage of the vehicle.

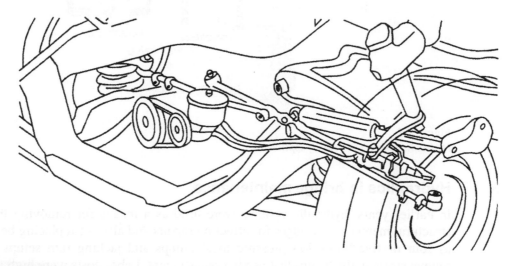

Figure 9.4
Bendix linkage-type power steering

The control valve of the two-unit type installation is mounted in one of the ball joints, usually at the Pitman arm. A small movement in the valve serves to open and close the hydraulic ports and thus operate the double acting power cylinder. Installation of the power cylinder and control valve can be made without changing the existing geometry of the steering linkage.

In effect, the existing steering system including the steering gear remains intact. Likewise the system is free to operate entirely by physical effort, when the engine is not running and in the absence of hydraulic pressure.

9.5.4 Welding application

A hydraulic system can be used for holding and positioning the parts to be welded during a welding operation. It is a typical example of how fluid power can be used in manufacturing and production operations, to reduce the overall costs and to increase production.

This application requires a sequencing system for fast and positive holding. This is accomplished by placing a restrictor (sequence valve) in the line leading to the second cylinder, as shown in Figure 9.5. The first cylinder extends to the end of its stroke.

Oil pressure then builds up, overcoming the restrictor setting and the second cylinder extends to complete the 'hold' cycle. This unique welding application of hydraulics was initiated to increase productivity.

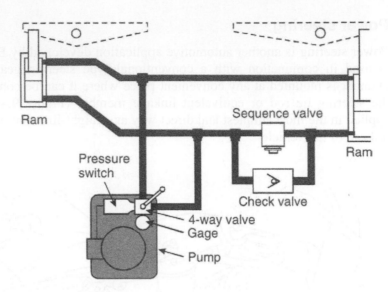

Figure 9.5
Hydraulic circuit for welding application

9.5.5   Hydraulics in bridge maintenance

In earlier years, hydraulic systems were used as a means for removing the stress from structural members of bridges for effecting repairs and also for replacing beams. As many as four or five bulky low-pressure hand pumps and jacking ram setups were used to relieve stress in the beams that needed replacement. Labor costs were high and there were no accurate means of recording the pressures.

In recent times, modern hydraulic systems have been designed, with capacities to locate several 100 tons rams on the bridge structure. Only one portable pump is used to actuate all the rams by the use of a special manifold, thereby simplifying the operation and making it easier to remove the stress from the members that need replacement.

10

Hydraulic circuit design and analysis

10.1 Objectives

After reading this chapter the student will be able to:

- Identify all the symbols used in hydraulic schematics
- Understand various hydraulic circuits
- Understand and explain hydraulic schematics effectively
- Design a hydraulic circuit for performing a desired function
- Analyze the function of each hydraulic circuit in an application.

10.2 Introduction

In the previous chapters, we have dealt with the basic fundamentals of hydraulics, hydraulic system components and their applications. Let us now discuss about hydraulic schematics and hydraulic circuits and the various terminologies associated with them.

   As we have seen earlier, a hydraulic circuit comprises a group of components such as pumps, actuators, control valves and conductors arranged to perform a useful task. When analyzing or designing a hydraulic circuit, the following considerations must be taken into account:

- Safety of operation
- Performance of the desired function
- Efficiency of operation.

10.3 Symbols of hydraulic components

It is very important for a fluid power technician or a designer to have knowledge of each of the hydraulic components and their functions in a hydraulic circuit. Hydraulic circuits are developed by using graphical symbols for all of the components. Therefore it is pertinent to know the symbols of each and every component used in a hydraulic system.

   The symbols discussed here conform to the American National Standard Institute (ANSI) standards and are tabulated below (see Figure 10.1).

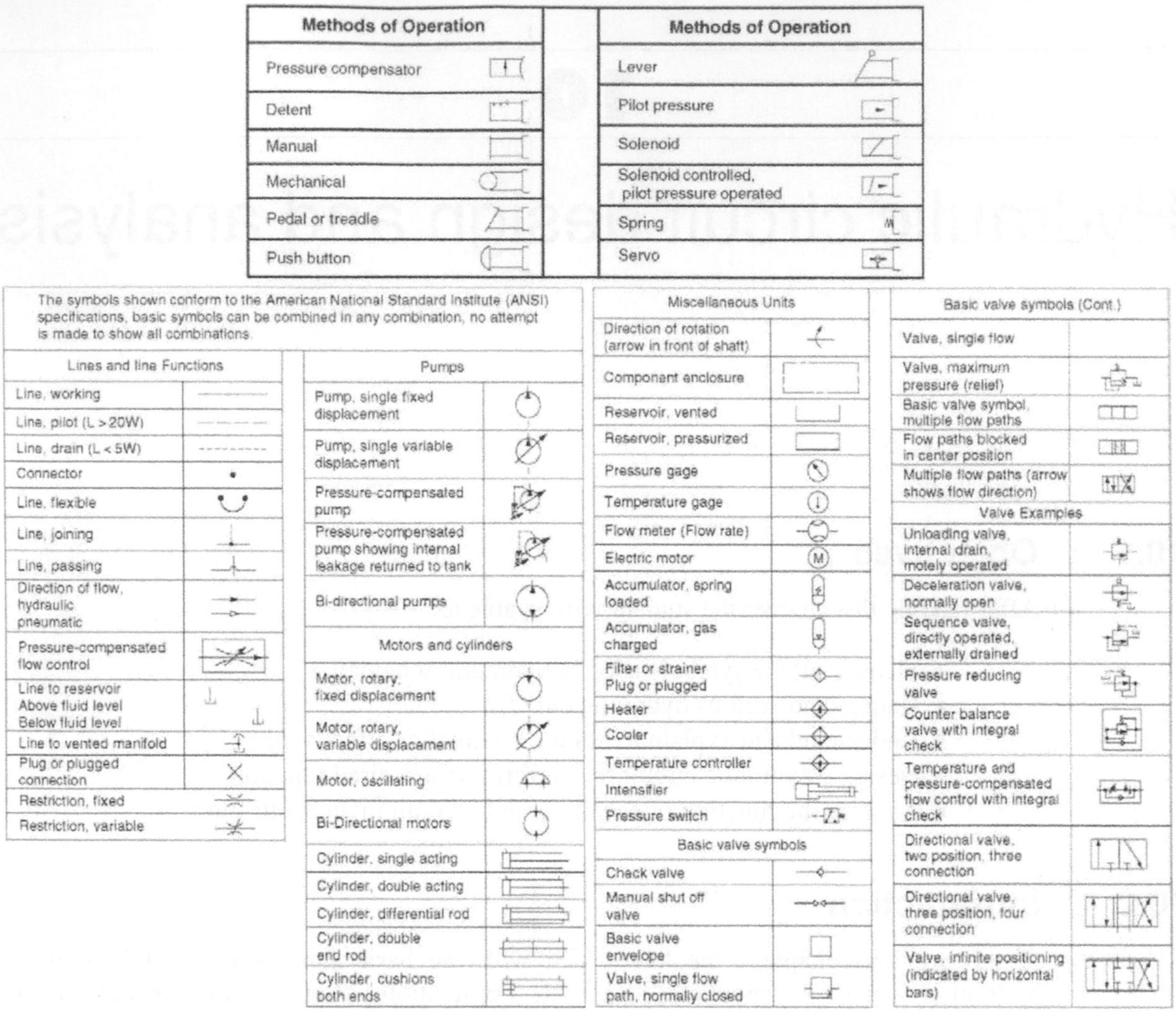

Figure 10.1
ANSI symbols for hydraulic components

10.4 Hydraulic circuits

In this section we shall take a look at how various types of hydraulic circuits are designed for efficient operation. We shall examine the following circuits:

- Control of a double acting hydraulic cylinder
- Regenerative circuit
- Pump unloading circuit
- Counterbalance valve application
- Hydraulic cylinder sequencing circuit
- Fail-safe circuits
- Speed control of a hydraulic motor
- Mechanical hydraulic servo system.

10.4.1 Control of a double acting hydraulic cylinder circuit

This circuit is designed as shown in Figure 10.2.

When the four-way valve is in its spring-centered position, the cylinder is hydraulically locked. Also the pump is loaded back to the tank at atmospheric pressure.

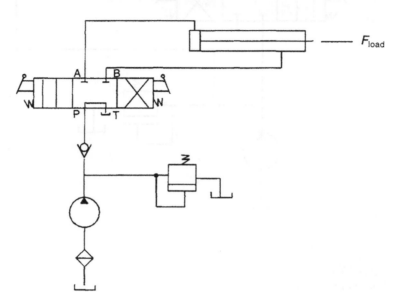

Figure 10.2
Control of a double acting hydraulic cylinder

When the four-way valve is actuated into the flow path configuration of the left envelope, the cylinder is extended against its force load (F_{load}) as oil flows from port P through port A. The oil at the rod end of the cylinder is free to flow back into the reservoir through the four-way valve from port B through port T. The cylinder will not extend if the oil in the rod end is not allowed to flow back to the reservoir.

When the four-way valve is de-activated, the spring-centered envelope prevails, and the cylinder is once again hydraulically locked.

When the four-way valve is actuated in the right envelope configuration, the cylinder retracts, as oil flows from port P through port B. Oil in the blank end is allowed to flow back to the reservoir from port A through port T of the four-way valve. At the end of the stroke, there is no system demand for oil. Therefore the pump flow goes through the relief valve at its set pressure, unless the four-way valve is de-activated. In any event, the system is protected from cylinder overloads.

The check valve prevents the load from retracting the cylinder, while it is being extended using the left envelope flow path configuration.

10.4.2 Regenerative circuit

Figure 10.3 shows a regenerative circuit used to accelerate the extending speed of the double acting hydraulic cylinder.

In this system, both ends of the hydraulic cylinder are connected in parallel and one of the ports of the four-way valve is blocked. The operation of the cylinder during the retraction stroke is the same as that of a regular double acting cylinder.

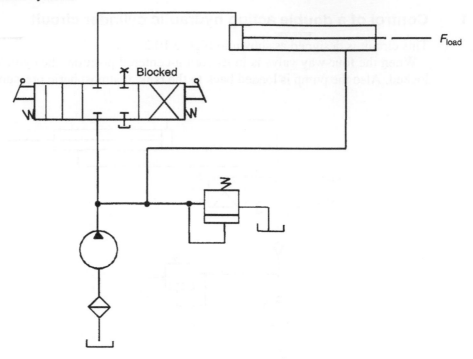

Figure 10.3
Regenerative circuit

Fluid flows through the DCV (direction control valve) via the right envelope during the retraction stroke. In this mode the fluid from the pump bypasses the DCV and enters the rod end of the cylinder. Fluid in the blank end drains back to the tank through the DCV as the cylinder retracts.

When the DCV is shifted to the left envelope configuration, the cylinder extends. The speed of extension is greater than that for a regular double acting cylinder. This is because the flow from the rod end (Q_R) regenerates with the pump flow (Q_P) to provide a total flow rate (Q_T), which is greater than the pump flow rate to the blank end of the cylinder.

The equation for extending speed can be obtained as follows:

The total flow rate entering the blank end of the cylinder equals the pump flow rate plus the regenerative flow rate coming from the rod end of the cylinder.

$$Q_T = Q_P + Q_R$$

The regenerative flow rate equals the difference of the piston and rod areas ($A_P - A_R$), multiplied by the extending speed of the piston (V_{Pext}).

$$Q_P = A_P V_{Pext} - \left(A_P - A_R \right) V_{Pext}$$

Solving for the extending speed of the piston, we have

$$V_{Pext} = \frac{Q_P}{A_R}$$

The retracting speed of the piston equals the pump flow divided by the difference of the piston and rod areas.

$$V_{Pret} = \frac{Q_P}{(A_P - A_R)}$$

It should also be borne in mind that the load-carrying capacity of a regenerative cylinder during its extension stroke is less than that obtained for a regular double acting cylinder.

10.4.3   Pump unloading circuit

Figure 10.4 depicts a circuit used for unloading a pump using an unloading valve.

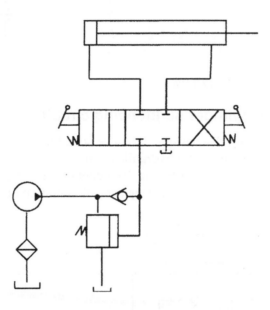

Figure 10.4
Pump unloading circuit

In this circuit, the unloading valve opens as the cylinder reaches the end of its extension stroke. This is because the check valve keeps the high-pressure oil in the pilot line of the unloading valve. When the DCV is shifted to retract the cylinder, the motion of the cylinder reduces the pressure in the pilot line of the unloading valve. This resets the unloading valve until the cylinder is fully retracted at the point where the unloading valve unloads the pump. It is thus seen that the unloading valve unloads the pump at the end of the extending and retracting strokes as well as in the spring-centered position of the DCV.

10.4.4   Counterbalance valve application

Figure 10.5 illustrates the use of a counterbalance or backpressure valve to keep a vertically mounted cylinder in upward position during pump idling.

The counterbalance valve is set to open at a pressure slightly above the pressure required to hold the piston up. This permits the cylinder to be forced downward, when pressure is applied on the top. The open center direction control valve unloads the pump. The DCV used here is a solenoid-actuated, spring-centered valve with an open center flow path configuration.

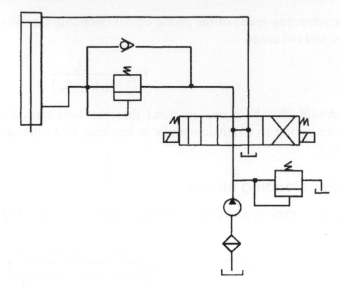

Figure 10.5
Counterbalance valve application

10.4.5   Hydraulic cylinder sequencing circuit

From our earlier discussions, we have seen how a sequence valve can be used to create sequential operations in a hydraulic circuit.

The circuit depicted in Figure 10.6 contains a hydraulic system in which two sequence valves are used to control the sequence of operation of two double-acting cylinders.

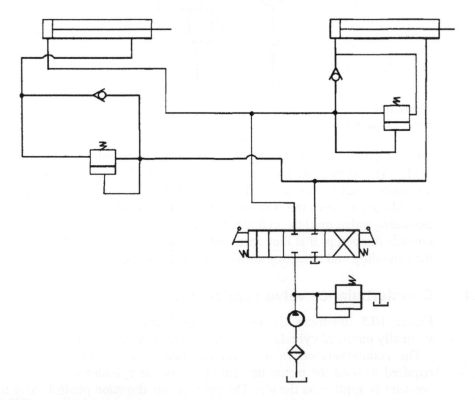

Figure 10.6
Hydraulic cylinder sequence circuit

When the DCV is shifted into its left envelope mode, the left cylinder extends completely and then the right cylinder extends. If the DCV is shifted into its right envelope mode, the right cylinder retracts fully followed by the left cylinder. This sequence of the cylinder operation is controlled by the sequence valves. The spring-centered position of the DCV locks both the cylinders in place.

The best example of this circuit is the case of a production operation. The left cylinder should extend in order to accomplish the job of clamping a work piece with the help of a power vice jaw. The right cylinder extends to drive a spindle to drill a hole in the work piece. After the hole has been drilled, the right cylinder retracts first and then the left one. The sequence valve installed in the circuit ensures that these operations occur in a predefined fashion.

10.4.6   Fail-safe circuit

Fail-safe circuits are basically designed to prevent injury to the operator or damage to the equipment. In general they prevent any accidental fall or overloading of the equipment.

Figure 10.7 shows a fail-safe circuit in which the cylinder is prevented from falling in the event of a hydraulic line rupture.

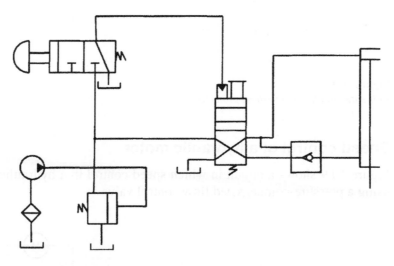

Figure 10.7
Fail-safe circuit for preventing cylinder fall in the event of hydraulic line rupture

To lower the cylinder, the pilot pressure from the blank end of the piston must pilot-open the check valve at the rod end, in order to allow the oil to return to the reservoir through the DCV. This happens when the push button valve is actuated to permit pilot pressure actuation of the DCV or with direct manual operation of the DCV during pump operation. The pilot-operated DCV allows free flow in the opposite direction to retract the cylinder when this DCV returns to its spring offset mode.

Figure 10.8 is another example of a fail safe circuit in which overload protection is provided for the system components.

The direction control valve 1 is controlled by a push button three-way valve 2. When overload valve 3 is in its spring-offset mode, it drains the pilot line of valve 1. If the cylinder experiences excessive resistance during its extension stroke, the sequence valve 4 pilot-actuates overload valve 3. This drains the pilot line of valve 1, causing it to return to its spring-offset mode. If push button valve 2 is then operated, nothing will happen unless the overload valve 3 is manually shifted into its blocked port configuration. This ensures the protection of the system components against excessive pressure due to excessive cylinder load during the extension stroke.

The best example of this circuit's use case of a production operation: the left cylinder should extend in order to accomplish the job of clamping a work piece with the help of a vise jaw. The right cylinder extends in order to enable a spindle to drill a hole in the work piece. A sequence valve is used to ensure that the right cylinder advances first and then the left one.

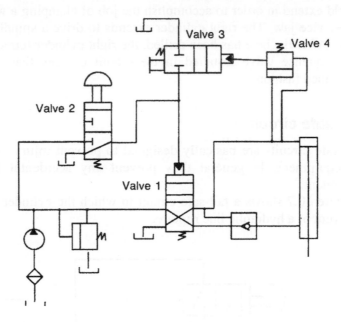

Figure 10.8
Fail-safe circuit with overload protection

10.4.7   Speed control of a hydraulic motor

Figure 10.9 shows a circuit in which speed control in a hydraulic circuit is accomplished using a pressure-compensated flow control valve.

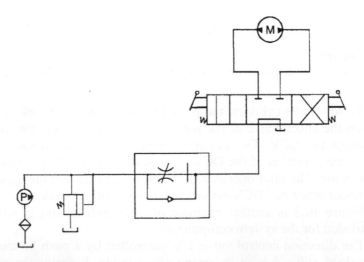

Figure 10.9
Speed control of a hydraulic motor

The operation of the circuit is as follows:

- In the spring-centered position of the tandem four-way valve, the motor is hydraulically locked.
- When the four-way valve is actuated into the left envelope, the motor rotates in one direction. Its speed can be varied by adjusting the setting of the throttle of the flow control valve. The speed can be infinitely varied as the excess oil goes through the pressure relief valve.
- When the four-way valve is de-activated, the motor stops suddenly and gets locked.
- When the right envelope of the four-way valve is in operation, the motor rotates in the opposite direction. The pressure relief valve provides overload protection when the motor experiences an excessive torque load.

10.4.8   Mechanical hydraulic servo system

Figure 10.10 shows a mechanical hydraulic servo system with automotive power steering, the sequential operation of which occurs as follows:

- The input or command signal is the turning of the steering wheel.
- This results in movement of the valve sleeve, which ports oil to the actuator (steering cylinder).

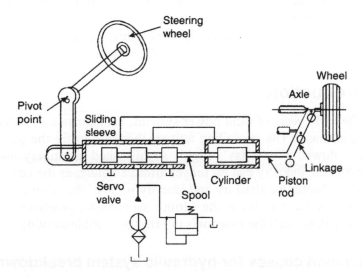

Figure 10.10
Mechanical hydraulic servo system

- The piston rod moves the wheels through the steering linkage.
- The valve spool is attached to the linkage, thereby moving it.

When the valve spool has moved far enough, it cuts off the oil flow through the cylinder. This stops the motion of the actuator.

It is therefore clear that mechanical feedback re-centers the valve (servo valve) in order to stop motion at the desired point which in turn is determined by the position of the steering wheel. Additional motion of the steering wheel is required to cause further motion of the output wheels.

11

Maintenance and troubleshooting

11.1 Objectives

After reading this chapter the student will be able to:

- Understand and explain the various causes for failure in a hydraulic system
- Carry out preventive maintenance of system
- Understand the functions and importance of sealing devices in a hydraulic system
- Carry out preliminary troubleshooting activities for determining the causes of malfunctioning in a hydraulic system.

11.2 Introduction

In the early years of fluid power systems, maintenance was frequently performed on a hit or miss basis. The prevailing attitude then was to fix the problem only after the system broke down. With today's highly sophisticated machinery and with the advent of mass production, industry can no longer afford a failure, as the cost of downtime is prohibitive. In this chapter we shall try and identify some of the common causes of hydraulic system failures and also examine the various maintenance practices to be followed in a hydraulic system along with the essentials of effective troubleshooting.

11.2.1 Common causes for hydraulic system breakdown

The most common causes of hydraulic system failures are:

- Clogged and dirty oil filters
- An inadequate supply of oil in the reservoir
- Leaking seals
- Loose inlet lines, which cause pump cavitations and eventual pump damage
- Incorrect type of oil
- Excessive oil temperature
- Excessive oil pressure.

A majority of these problems can be overcome through a planned preventive maintenance regime. The overall design of the system is another crucial aspect. Each component in the system must be properly sized, compatible with, and form an integral part of the system.

It is also imperative that easy access be provided to components requiring periodic inspection and maintenance such as strainers, filters, sight gages, fill and drain plugs and the various temperature and pressure gages. All hydraulic lines must be free of restrictive bends, as this tends to result in pressure loss in the line itself.

The three maintenance procedures that have the greatest effect on system life, performance and efficiency are:

1. Maintaining an adequate quantity of clean and proper hydraulic fluid with the correct viscosity
2. Periodic cleaning and changing of all filters and strainers
3. Keeping air out of the system by ensuring tight connections.

A vast majority of the problems encountered in hydraulic systems have been traced to the hydraulic fluid, which makes frequent sampling and testing of the fluid, a vital necessity. Properties such as viscosity, specific gravity, acidity, water content, contaminant level and bulk modulus require to be tested periodically. Another area of vital importance is the training imparted to maintenance personnel to recognize early symptoms of failure. Records should also be maintained of past failures and the maintenance action initiated along with data containing details such as oil tests, oil changes, filter replacements, etc.

Oxidation and corrosion are phenomena which seriously hamper the functioning of the hydraulic fluid. Oxidation which is caused by a chemical reaction between the oxygen present in the air and the particles present in the fluid, can end up reducing the life of the fluid quite substantially. A majority of the products of oxidation are acidic in nature and also soluble in the fluid, thereby causing the various components to corrode.

Although rust and corrosion are two distinct phenomena, they both contribute a great deal to contamination and wear. Rust, which is a chemical reaction between iron and oxygen, occurs on account of the presence of moisture-carrying oxygen. Corrosion on the other hand is a chemical reaction between a metal and acid. Corrosion and rust have a tendency to eat away the hydraulic component material, causing malfunctioning and excessive leakage.

11.2.2 The phenomenon of wear due to fluid contamination

Excessive contaminants in the working fluid prevent proper lubrication of components such as pumps, motors, valves and actuators. This can result in wear and scoring which affect the performance and life of these components and leads to their eventual failure. A typical example of this is the scored piston seal and cylinder bore of cylinders causing severe internal leakage and resulting in premature cylinder failure.

11.2.3 Problems due to entrained gas in fluids

Entrained gas or gas bubbles in the hydraulic fluid is caused by the sweeping of air out of a free air pocket by the flowing fluid and also when pressure drops below the vapor pressure of the fluid. Vapor pressure is that pressure at which the fluid begins changing into vapor. This vapor pressure increases with increase in temperature. This results in the creation of fluid vapor within the fluid stream and can in turn lead to cavitation problems in pumps and valves. The presence of these entrained gases reduces the effective bulk modulus of the fluid causing unstable operation of the actuators.

The phenomenon of cavitation is in fact the formation and subsequent collapse of the vapor bubbles. This collapse of the vapor bubbles takes place when they are exposed to the high-pressure conditions at the pump outlet, creating very high local fluid velocities, which impact on the internal surfaces of the pump. These high-impact forces cause flaking or pitting on the surface of components such as gear teeth, vanes and pistons leading to premature pump failure. Additionally the tiny metal particles tend to enter and damage other components in the hydraulic system. Cavitation can also result in increased wear on account of the reduced lubrication capacity.

Cavitation is indicated by a loud pump noise and also by a decreased flow rate as a result of which the pressure becomes erratic. Air also tends to get trapped in the pump line due to a leak in the suction or on account of a damaged shaft seal. Additionally it has to be also ensured that air escapes through the breather while the fluid is in the reservoir or otherwise it tends to enter the pump suction line. To counter the phenomenon of cavitation in pumps, the following steps are recommended by manufacturers:

1. Suction velocities to be kept below 1.5 m/s (5 ft/s)
2. Pump inlet lines to be kept as short as possible
3. Pump to be mounted as close to the reservoir as possible
4. Low-pressure drop filters to be used in the suction line
5. Use of a properly designed reservoir that will help remove the trapped air in the fluid
6. Use of hydraulic fluid as recommended by the manufacturer
7. Maintaining the oil temperature within prescribed limits, i.e. around 65 °C or 150 °F.

11.3   Safety

Electrical systems are generally recognized as being potentially lethal and all organizations by law must have procedures for isolation of the equipment and adopt safe working practices. Hydraulic and pneumatics are no less dangerous but tend to be approached in a far more laissez faire or casual manner.

A hydraulic system can present the following dangers to an operator:

- High-pressure air or oil released suddenly can attain explosive velocities and can easily cause an accident.
- The unexpected movement or drift of components such as cylinders can be harmful.
- Spilt hydraulic oil is very slippery and can cause accidents.

A few guidelines to ensure safety in hydraulic systems are listed here:

- The implications arising out of any action have to be considered before resorting to it.
- Anything that can move with change in pressure as a result of your actions should be mechanically secured or guarded.
- Particular care should be taken with regard to suspended loads. It must be remembered that fail-open valves will turn ON when the system is de-pressurized.
- Never disconnect pressurized lines or components. The whole system should be de-pressurized before disconnecting any of the lines.
- Put up safety notices to prohibit operation by other people.
- Ensure that the accumulators in the hydraulic system are fully blown down.
- Make proper arrangements to prevent spillage of oil on the floor.

- Where there is an electrical interface to a hydraulic system (e.g., solenoids, pressure switches, limit switches) the control circuit should be isolated, not only to reduce the risk of electric shock but also to reduce the possibility of fire.
- After the work is completed, keep the area tidy and clean. Check for any leakages and confirm correct operation of the system.
- Many components contain springs under pressure. If released in an uncontrolled manner, these can fly out at high speed and cause injury. Springs should be removed with utmost care.

In USA, the Occupational Safety and Health Administration (OSHA) of the Department of Labor describes and enforces safety standards at industry locations where the hydraulic equipment is operated. For detailed information of OSHA standards and requirements, the OSHA publication 2072 can be referred to. The general industry guide for applying safety and health standards, 29 CFR 1910 also provides us with a standard set of safety standards for operating hydraulic equipment.

These standards deal with the following categories:

- Workplace standards
- Machines and equipment standards
- Materials standards
- Employee standards
- Power source standards
- Process standards.

The basic rule to follow is that there should be no compromise when it comes to the health and safety of people at the place of their work.

11.4   Cleanliness

Most hydraulic and pneumatic faults are caused by dirt. Very small particles can nick seals, abrade surfaces, block orifices and cause valve spools to jam. Ideally components should not be dismantled in the usual dirty conditions found on site. These should be taken to a clean workshop equipped with proper workbenches. Components and hoses come from manufacturers with their orifices sealed with plastic plugs, to prevent dirt ingress during transit. These should be left as they are during storage and removed only when the component is to be put to use.

Filters are meant to remove dirt particles, but only work until they are clogged. A dirty filter may cause the fluid to bypass and can make things far more worse by accumulating the particles and then releasing them all in one lump. Filters should be regularly checked and cleaned or changed when required.

The oil condition in a hydraulic system is also crucial for maintaining reliability. Oil, which is dirty, oxidized or contaminated, forms a sticky gummy sludge which blocks small orifices and causes the valve spools to jam. The oil condition should be regularly checked and the suspect oil changed before problems develop.

11.5   Preventive maintenance

Most of the production personnel carry the impression that a maintenance department exists primarily to repair the faults that occur. Unfortunately this is not the case. The most important part of the maintenance department's responsibility is to perform routine planned maintenance otherwise known as preventive maintenance.

Preventive maintenance primarily deals with:

- Regular servicing of the equipment
- Checking for correct operation
- Identification of potential faults and their immediate rectification or correction.

As an often-overlooked side benefit, planned maintenance trains the maintenance technician in the proper operation and layout of the plant for which they are responsible. Most of the common problems listed in the introductory section of this chapter can be eliminated if a planned preventive maintenance program is undertaken.

More than 50% of the problems encountered in hydraulic systems have been observed to be related to hydraulic oil. This is why regular sampling and testing of the hydraulic fluid is a very important. A portable hydraulic fluid test kit is available nowadays. This helps in carrying out the basic tests at the site itself. Tests that can be performed include ones such as determination of viscosity, water content and particulate contamination.

It is vital that the maintenance personnel be trained to carry out maintenance activities effectively. A technician should also be able to recognize the early symptoms of potential hydraulic problems. For example, a noisy pump may be due to cavitation caused by a clogged inlet filter. It might also be due to a loose inlet fitting which permits air ingress into the pump. If the cavitation is due to air leakage in the pump, the oil in the reservoir tends to get covered with foam. When air becomes entrained in the oil, it causes spongy operation of the hydraulic actuators. A sluggish actuator may also be due to the high viscosity of the fluid.

For preventive maintenance techniques to be really effective, it is necessary to have a good reporting and recording system. These reports should include the following:

- The type of symptoms encountered and how they were detected along with the respective date
- A description of the maintenance repairs performed. This should include the replacement of parts, the amount of downtime and the date
- Records of dates when the oil was tested, added or changed
- Records of dates when the filters were cleaned or replaced.

Proper maintenance procedures with respect to external oil leakages are also essential. Safety hazards due to oil spillage on the floor should be prevented. The bolts and brackets of loose mountings should be tightened as soon as they are detected as they can cause misalignment of the pump and actuator shafts.

11.5.1 Sealing devices

Oil leakage in a hydraulic system reduces efficiency and leads to increased power losses. Internal leakage does not cause loss of fluid in the system as the fluid returns to the reservoir. External leakage represents a loss of fluid in the system. An improperly assembled pipe fitting is the most common cause of external leakage. Shaft seals on pumps and cylinders can get damaged due to misalignment, leading to leakages in the system.

Seals are used in hydraulic equipment to prevent excessive internal and external leakages and to keep out contamination. Seals can be of a positive or non-positive type and are generally designed for static or dynamic applications. Positive seals do not allow any leakage whatsoever. Non-positive seals permit a small amount of internal leakage.

Static seals are used between mating parts, which do not move relative to each other. Figure 11.1 shows some of the static seals including flanged gaskets and seals. The static seals are compressed between two rigidly connected parts. They represent a simple and non-wearing joint, which would be trouble free if properly assembled.

Dynamic seals are assembled between mating parts, which move relative to each other. Dynamic seals are subject to wear and tear as one of the mating parts rubs against the seal. The most widely used seals of this type are:

- O-rings
- Compression packings
- Piston cup packings
- Piston rings
- Wiper rings.

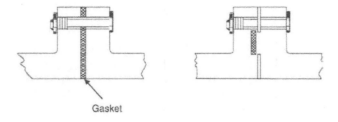

Gasket

Metal-to-metal joints V

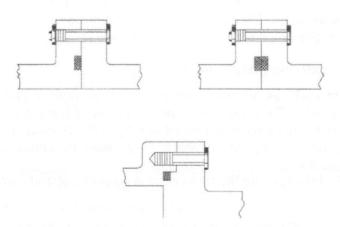

Figure 11.1
Static seal flange joints

V-ring packings

V-ring packings are compressible type seals, which are used in virtually all reciprocating motion applications. These include rod and piston cylinders in hydraulic cylinder applications, press rams and jacks. Here, proper adjustment is essential since excessive tightening will accelerate wear and tear.

Piston cup packings

Piston cup packings are designed specifically for pistons in reciprocating pumps and hydraulic cylinders. They offer the best service life for this type of application. Figure 11.2 shows a typical installation of piston cup rings for double acting and single acting operations.

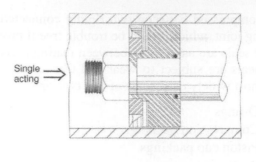

Single
acting ⟹

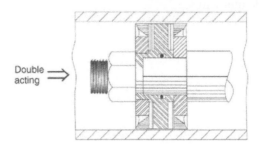

Double
acting ⟹

Figure 11.2
Typical application of piston cup packings

Sealing is accomplished when the pressure pushes the cup lip outwards against the cylinder barrel. The backing plate and the retainers clamp the cup packing tightly in its place, allowing it to handle very high pressures.

Non-metallic piston ring packings

These packings are made out of tetrafluoro ethylene (TFE), a chemically inert, tough waxy solid. Their extremely low coefficient of friction permits them to run completely dry and at the same time prevent scoring of the cylinder walls. This type of piston ring is very ideal for applications where the presence of lubrication can be detrimental or even dangerous.

The following are the most common types of materials used for seals:

- *Leather*: This is rugged and inexpensive. However it tends to 'squeal' when dry and cannot operate above 93 °C (200 °F). Leather also does not operate well at cold temperatures of around –50 °C (–58 °F).
- *Buna-N*: This material is rugged, inexpensive and wears well. It has a wide operating range between –45 °C (–50 °F) and 121 °C (250 °F) and also maintains good sealing characteristics in this range.
- *Silicone*: This elastomer has an extremely wide temperature range between –68 °C (–90 °F) and 232 °C (450 °F). Hence it is widely used in rotating shaft seals and static seals. Silicone is not used for reciprocating seal application as it has a low resistance to tear.
- *Neoprene*: This material has a temperature range of –54 °C (–65 °F) to 120 °C (250 °F). It has a tendency to vulcanize beyond this temperature.
- *Viton*: This material contains about 65% fluorine. It has become a standard material for elastomer type seals for use at elevated temperatures up to 260 °C (500 °F). The minimum temperature at which these seals operate is about –29 °C (–20 °F).

- *Tetrafluoroethylene*: It is a form of plastic and is a very widely used seal material. It is quite tough and chemically inert in nature and has excellent resistance up to temperatures of 370 °C (around 700 °F). It additionally possesses an extremely low coefficient of friction. One major disadvantage with this material is its tendency to flow under pressure forming thin films. This can be neutralized to a large extent by using filler materials such as graphite, asbestos and glass fibers.

11.6 Troubleshooting

Troubleshooting or faultfinding as we call it, is often performed in a random and haphazard manner, leading to items being changed for no logical reason. Such an approach may eventually work but it is hardly the quickest and cheapest way of getting a faulty system back into operation. There must be better and more systematic approaches to correcting a problem. Fault finding has been rather simplistically, represented in the form of a flow chart below (Figure 11.3).

11.6.1 Fault finding process

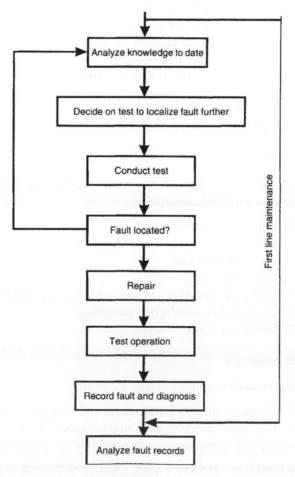

Figure 11.3
Flow chart depicting the process of fault finding

Recognizing trouble indications

Before going into the specifics of fault finding, let us discuss a few general indications of malfunctioning in a hydraulic system.

Excessive heat

Excessive heat in a hydraulic system is a serious indication of trouble, the causes for which could be the following:

1. A misaligned coupling that results in excessive load on the bearings generating heat.
2. An unusually warmer return line that could be due to operation at relief valve setting.
3. The phenomena of cavitation and slippage in pumps will also generate excessive amount of heat.
4. Increase in the internal leakage of components due to the use of low-viscosity hydraulic fluids and leading to heat generation.

Excessive noise

Excessive noise can result from:

1. Problems related to wear and misalignment
2. Pump cavitation or the presence of air in the hydraulic fluid
3. Presence of contaminants in the hydraulic fluid that may cause a relief valve to chatter and produce noise
4. Low reservoir levels and contaminated filters
5. Excessive drive speed and loose intake lines
6. High fluid viscosity and
7. Damaged or worn couplings.

Incorrect flow

The following are a list of common causes that might lead to incorrect pressure conditions in a hydraulic circuit:

1. Improper reservoir level
2. Dirty, clogged filters and strainers
3. Clogged inlet line
4. Defective pump
5. Leaky connections and the presence of air in the system
6. Damaged or misaligned couplings
7. Defective control valves.

Incorrect pressure

Incorrect pressure conditions can result from:

1. Contaminated hydraulic fluid and clogged filters
2. Improper reservoir level and presence of air.

When troubleshooting hydraulic systems, it should be kept in mind that the pump produces fluid flow. However there must be resistance to flow in order to have a pressure. The following is a list of hydraulic system operating problems and the probable causes, which should be investigated during troubleshooting.

Faults	**Probable Causes**
1. Noisy pump	(a) Air entering the pump inlet
	(b) Misalignment of the pump
	(c) Excessive oil viscosity
	(d) Dirty inlet strainer
	(e) Chattering relief valve
	(d) Damaged pump
	(f) Excessive pump speed
	(g) Loose or damaged inlet.
2. Low or erratic pressure	(a) Air in the fluid
	(b) Pressure relief valve set too low
	(c) Pressure relief valve not properly seated
	(d) Leak in the hydraulic line
	(e) Defective or worn-out pump
	(f) Defective or worn-out actuator.
3. No pressure	(a) Pump rotating in the wrong direction
	(b) Ruptured hydraulic line
	(c) Low oil level in the reservoir
	(d) Pressure relief valve malfunctioning
	(e) Full pump flow by-passed to the tank due to faulty valve.
4. Actuator fails to move	(a) Faulty pump
	(b) Direction control valve fails to shift
	(c) System pressure too low
	(d) Defective actuator
	(e) Pressure relief valve stuck open
	(f) Actuator load is excessive
	(g) Check valve installed in the reverse direction.
5. Slow or erratic motion of the actuator	(a) Air in the system
	(b) High viscosity of the fluid
	(c) Worn or damaged pump
	(d) Pump speed too low
	(e) Excessive leakage through actuators
	(f) Faulty or dirty flow control valves
	(g) Blocked air breather in the reservoir
	(h) Low fluid level in the reservoir
	(i) Faulty check valve
	(j) Defective pressure relief valve.
6. Overheating of hydraulic fluid	(a) Heat exchanger turned off or clogged
	(b) Undersized components or piping
	(c) Incorrect fluid
	(d) Continuous operation of pressure relief valve
	(e) Overloaded system
	(f) Dirty fluid
	(g) Reservoir too small
	(h) Inadequate supply of oil in the reservoir
	(i) Excessive pump speed
	(j) Clogged or inadequate sized air breather.

Now that we have understood the basic troubleshooting concepts with regard to individual components in a hydraulic system, let us get familiarized with troubleshooting techniques listed under five main categories. The effects indicating the system malfunctioning are in the form of flow charts, while the remedies are listed in steps, to help facilitate easy understanding.

1. *Excessive noise*

Flow chart (Figure 11.4):

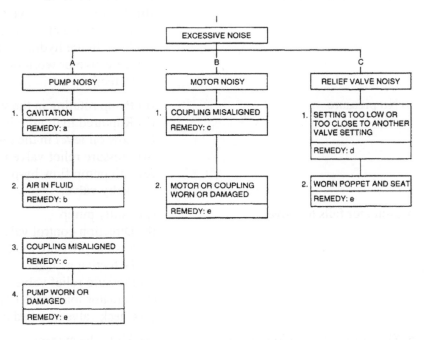

Figure 11.4
Trouble shooting for excessive noise

Remedies:

(a) Any or all of the following:

    1. Replace dirty filters and clean strainers, clogged inlet line and reservoir breather
    2. Change fluid, switch over to proper pump speed and overhaul/replace supercharger pump

(b) Any or all of the following:

    1. Fill reservoir to required level, bleed air from the system and plug leakages in inlet line
    2. Replace pump shaft seal/shaft

(c) Condition of bearings, seals and couplings to be checked and the unit aligned
(d) Pressures to be corrected
(e) Overhaul or replace.

2. *Excessive heat*

Flow chart (Figure 11.5):

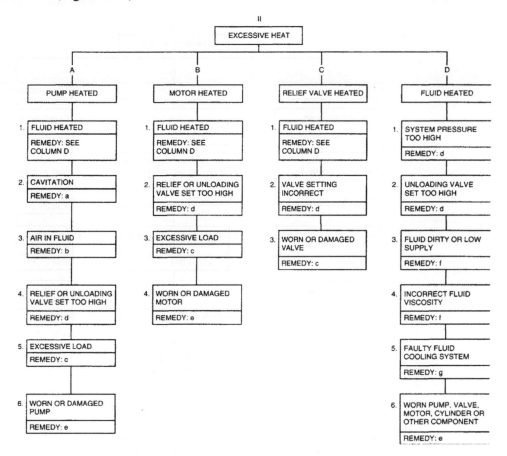

Figure 11.5
Trouble shooting for excessive heat

Remedies:

(a) Any or all of the following:

1. Replace dirty filters and clean clogged inlet line and reservoir breather
2. Change fluid, switch over to proper pump speed and overhaul/replace supercharger pump

(b) Any or all of the following:

1. Fill reservoir to required level, bleed air from the system and plug leakages in inlet line
2. Replace pump shaft seal/shaft

(c) Condition of bearings, seals and couplings to be checked and the unit aligned. Mechanical binding to be located and corrected

(d) Pressures to be corrected

(e) Overhaul or replace

(f) Filters to be replaced and hydraulic fluid to be changed in the event of improper fluid viscosity

(g) Clean cooler and cooler strainer, replace cooler control valve or repair or replace cooler.

3. *Incorrect flow*

Flow chart (Figure 11.6):

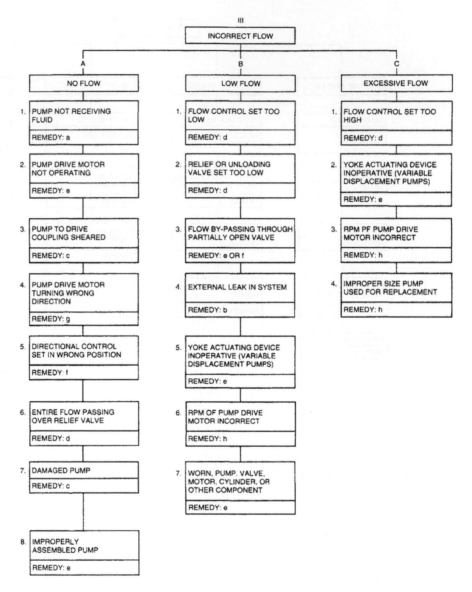

Figure 11.6
Trouble shooting for incorrect flow

Remedies:

(a) Any or all of the following:

1. Replace dirty filters and clean clogged inlet line and reservoir breather
2. Change fluid, switch over to proper pump speed and overhaul/replace supercharger pump

(b) Tighten leaky connections and carry out bleeding of the system

(c) Check for pump or pump drive damage. Replace and align coupling

(d) Adjust

(e) Overhaul or replace
(f) Check position of manually operated controls, check electrical circuit of solenoid controls. Repair or replace pilot pressure pump
(g) Reverse the direction of rotation
(h) Replace with correct unit.

4. *Incorrect pressure*

Flow chart (Figure 11.7):

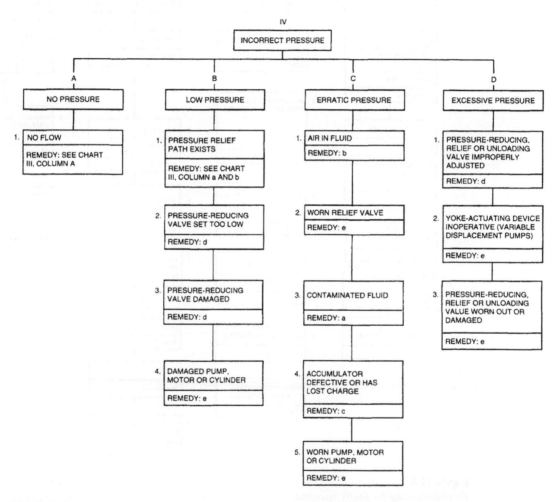

Figure 11.7
Trouble shooting for incorrect pressure

Remedies:

(a) Change dirty filters. Replace hydraulic fluid
(b) Tighten leaky connections, fill reservoir to proper level and carry out bleeding of the system
(c) Gas valve to be checked for leakage, charged to correct pressure and overhauled, if defective
(d) Adjust
(e) Overhaul or replace.

5. *Faulty operation*

Flow chart (Figure 11.8):

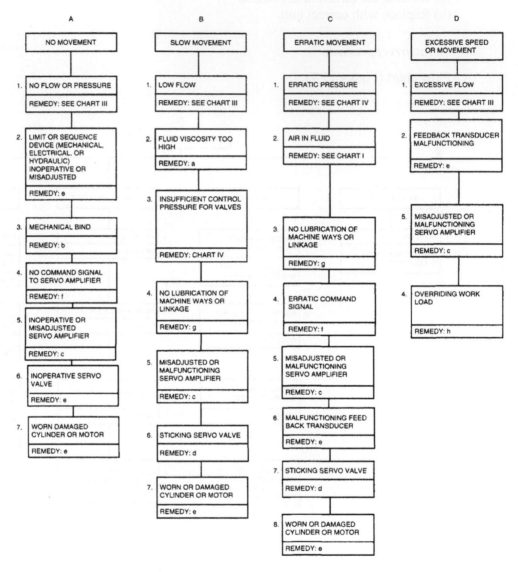

Figure 11.8
Troubleshooting for faulty operation

Remedies:

(a) Cold state of the fluid. Viscosity of the fluid also to be checked and if found improper, replaced with fluid having correct viscosity
(b) Locate bind and repair
(c) Adjust, repair or replace
(d) Clean, adjust or replace. Check system fluid condition and also condition of filters
(e) Overhaul or replace
(f) Repair command console/interconnecting wires
(g) Lubricate
(h) Adjust, repair or replace counterbalance valve.

Troubleshooting instruments

Hydraulic systems depend on proper flow and pressure from the pump to provide the necessary actuator motion for producing useful work. Hence the measurement of flow and pressure are two important means of troubleshooting a hydraulic system. Temperature is the third important parameter measured periodically as it affects the viscosity of oil. The use of flowmeters can help in determining whether or not the pump is producing the proper flow.

11.7 Commissioning procedures

Incorrect commissioning of hydraulic components during initial start-up can result in damage due to inadequate lubrication, cavitation and aeration that may not manifest itself for hundreds or even thousands of service hours. To avoid damage to the system components during start-up, the manufacturer's commissioning procedures should be followed wherever available.

The following are general procedures for commissioning hydraulic systems after components have been replaced from the system or if any other maintenance action has been carried out. The same procedures can be applied when commissioning new systems.

11.7.1 Pre-commissioning

If the system is down as a result of a major component failure, such as a pump failure:

- Drain and clean the reservoir to ensure that it is free from metallic debris and other contamination.
- Change all filters.
- Change the fluid. In large systems where the cost of changing the fluid may be prohibitive, the fluid should be circulated through a 10 μ filter (without bypass) before recharging it into the system.
- When fitting pumps and motors, check the drive coupling for its alignment with the pump shaft.
- On closed loop systems, closely inspect the high-pressure hoses or pipes and replace any suspect lines. A blown hose or a pipe can destroy a pump or a motor through cavitation.
- After fitting each cylinder, fill the cylinder with clean oil wherever possible, through its service port before connecting the service lines. This reduces the risk of air compression within the cylinder during start-up, which may result in damage to the seals, or the cylinder itself.
- After fitting motor and other connecting lines: In case of piston type motors, fill the motor casing with clean oil from the upper most port and connect the casing drain line.
- After fitting the pumps and connecting the service lines: Open the suction line valve at the reservoir and vent out all the air from the system at the pump suction line. This step is not necessary for a piston-type pump.

11.7.2 Commissioning

- Check that all the pipe and hose connections are tight.
- Confirm that the reservoir fluid level is above the minimum level.
- Confirm that all the controls are in neutral so that the system will start in an unloaded condition. Take safety precautions to prevent machine movement when the system is activated during initial start-up.

- Where the prime mover is an electrical equipment, confirm whether the direction of rotation of the motor is correct by inching the motor.
- Start the prime mover and run at the lowest possible speed (rpm).
- On closed loop systems, monitor the discharge pressure. If the manufacturer's specified charge pressure is not established within 20–30 s, shut down the prime mover and investigate the problem. Do not operate the system without adequate charge pressure.
- On variable displacement pumps and motors with external, low-pressure pilot lines, vent out the air from the pilot line and ensure that the line is full of oil. Caution! Do not bleed the pilot lines carrying high-pressure fluid – personal injury may result. If in doubt do not bleed pilot lines.
- Allow the system to run in idle mode and unloaded for 5 min. Monitor the pump for any unusual noises or vibrations and inspect the system for leaks and observe the reservoir fluid level.
- Operate the system without a load. Stroke the cylinders slowly, taking care not to develop pressure at the end of the stroke to avoid compression of trapped air, which may result in damage to the seals. Continue to operate in this manner until all the air is expelled and the actuators operate smoothly.
- With the system at the correct operating temperature, check and, if necessary, adjust pressure settings according to the manufacturer's specifications.
- Test the operation of the system with appropriate load.
- Inspect the system for leaks.
- Shut down the prime mover. Remove all gages fitted during commissioning and check the reservoir fluid level and top-up if necessary.
- Put the machine back into service.

11.8   Prevention of premature hydraulic component failure

Premature failure of hydraulic components decreases the productivity and increases the operating cost of a hydraulic system. This failure may simply be defined as the failure of a component prior to it achieving its expected service life. The expected life of individual components within the hydraulic system varies and is influenced by a number factors such as:

- Type of component
- Circuit design
- Operating load
- Duty cycle
- Operating conditions.

From an operation and maintenance perspective, the factor that has the most impact on component service life is the condition under which the hydraulic components operate. The following conditions will have a negative impact on hydraulic component service life and in extreme cases will lead to a premature failure.

11.8.1   High fluid temperature

A fluid temperature above 82 °C (180 °F) damages seals and reduces the life of the fluid. At higher temperatures, inadequate lubrication due to lower fluid viscosity causes damage to the system components. To avoid system damage due to overheating, it is important that a temperature alarm is fitted in the system.

11.8.2   Incorrect fluid viscosity

Generally, optimum operating efficiency is achieved with a fluid viscosity in the range of 16–36 cst. Maximum bearing life is achieved with a minimum viscosity of 25 cst. A very high fluid viscosity may damage the system components through cavitation, while low fluid viscosity may result in damage through inadequate lubrication.

11.8.3   Fluid contamination

Contamination of the hydraulic fluid may occur on account of the influence of air, water, solid particles or any other matter that impairs the function of a fluid.

Air contamination can result in damage to the system components through loss of lubrication, overheating and oxidation of seals. Common entry points for air contamination include the vortex effect at the pump suction (due to low reservoir oil level) or faulty seals. To avoid this, the reservoir oil level should always be maintained at the desired level.

Water contamination can result in damage to the system components through corrosion, cavitation and altered fluid viscosity. In order to avoid this, ensure that all possible points of penetration into the reservoir oil space are sealed. Also ensure that the maximum oil level is maintained, to minimize condensation within the reservoir.

Contamination from solid particles can result in damage to the system components through abrasive wear or can be generated internally. Common entry points of particle contamination are through the reservoir air space and on the surface of the cylinder rods.

To avoid this and to reduce the contamination load on the system's filters, the following measures should be undertaken:

- All penetration points into the reservoir airspace must be sealed and an air filter of 5 μ installed in the breather.
- The chrome surfaces of the cylinder rods must be made free from pitting, dents and scoring.
- The filters should be replaced regularly and fluid contamination levels monitored through regular sampling.

11.8.4   Incorrect commissioning or adjustment

Incorrect commissioning of hydraulic components can result in damage to the system components through inadequate lubrication, cavitation and aeration. Additionally, incorrect settings of the hydraulic system adjustments can result in component damage through over-pressurization, cavitation and aeration.

11.8.5   After the event

When premature failure does occur, a thorough investigation should be conducted, to understand the root cause of the failure. Consult a hydraulic specialist if necessary. Although, the failure analysis is not conclusive in all cases, it can provide valuable clues at identifying the cause of failure. This is essential in order to effect remedial action aimed at preventing a re-occurrence of the same.

Appendix A
Practical problems

Exercise 1

1.1 Calculate the pressure due to a column of 0.5 m of

   (a) Water
   (b) Oil of specific gravity 0.8
   (c) Mercury of specific gravity 13.6.

1.2 The intensity of pressure at a point in a fluid is 4 N/cm^2. Find the corresponding height of fluid when the fluid is

   (a) Water
   (b) Oil of specific gravity 0.9.

1.3 A bullet of mass 8 g travels at a speed of 400 m/s. It penetrates a target, which offers a constant resistance of 1000 N to the motion of the bullet. Calculate

   (a) The kinetic energy of the bullet
   (b) The distance through which the bullet has penetrated.

1.4 A hydraulic press has a ram of 25 cm diameter and a plunger of 4.0 cm diameter. Find the weight lifted by the hydraulic press when the force applied at the plunger is 400 N.

1.5 Calculate the gage pressure and absolute pressure at a point 3 m below the free surface of a liquid having a density of 1.53×10^3 kg/m^3, if the atmospheric pressure is equal to 750 mm of mercury. The specific gravity of mercury is 13.6 and density of water is 1000 Kg/m^3.

1.6 The diameters of a pipe are 15 and 20 cm at sections 1 and 2 respectively. The velocity of water flowing through the pipe is 4 m/s at section 1. Determine the discharge through the pipe and also the velocity of flow at section 2.

1.7 A pipe of 40 cm diameter carrying water, branches into two other pipes of diameters 25 and 20 cm respectively. If the average velocity in the 40 cm pipe is 2.0 m/s find the flow rate in this pipe. Also determine the velocity in the 20 cm pipe if the average velocity in the 25 cm pipe is 1.8 m/s.

Exercise 2

This exercise consists of problems involving designing and analyzing of simple hydraulic circuits, along with their solutions.

Sample Problem 1

Deals with air to hydraulic pressure booster system: The following exercise will tell us how to calculate the load-carrying capacity in a pressure booster system.

Problem:

The figure shows a pressure booster system used to drive a load 'F' via a hydraulic cylinder. The following data is available:

Inlet air pressure (P_1) = 100 psi
Air piston area (A_1) =   20 sq.in.
Oil piston area (A_2) =   1 sq.in.
Load piston area (A_3) =   25 sq.in.

Find the load-carrying capacity of the system.

Problem 1a

Solve the following problem on the same lines as above (problem to be solved by participants).

For the same system above, following data is given and you need to find the required load piston area.

Inlet air pressure (P_1) = 125 psi
Air piston area (A_1) = 20 sq.in.
Oil piston area (A_2) = 1 sq.in.
Load-carrying capacity (F) = 75 000 lb

Sample Problem 2

Deals with the hydraulic horsepower analysis technique.

Problem:

A hydraulic cylinder is used to compress a car body down to a bale size in 10 s. The operation requires a 10 ft stroke (S) and a force (F_{load}) of 8000 lb. If a 1000 psi pressure (P) pump is selected, find

    (a) The required piston area
    (b) The necessary pump flow rate
    (c) The hydraulic horsepower delivered by the cylinder.

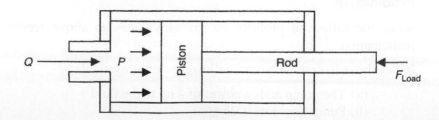

Problem 2a

Solve the following problem on the same lines as above (problem to be solved by participants).

A hydraulic cylinder is to compress a car body down to bale size in 8 s. The operation requires an 8 ft stroke and a 15 000 lb force. Find the following if the piston area is 12 sq. ins.

> (a) The pressure to be developed by the pump
> (b) The necessary flow rate of the pump
> (c) The hydraulic horsepower developed by the cylinder.

Sample Problem 3

In the following exercise, we shall study how a problem in a hydraulic system can be solved using the metric-SI system units.

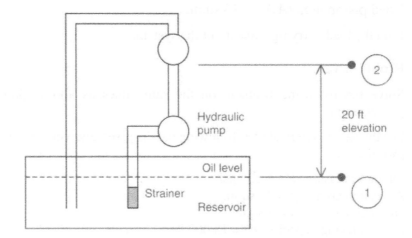

Problem:

For the above hydraulic system,

> 1. The pump adds a power of 5 hp (3730 W) to the fluid.
> 2. Pump flow rate is 0.001896 m³/s (6825.6 lph)
> 3. The pipe diameter is 25.4 mm (0.0254 m)
> 4. The specific gravity of oil is 0.9
> 5. The elevation difference between station 1 and 2 is 20 ft (6.096 m)
> 6. The head loss between the two stations is 9.144 m of oil.

Find the pressure available at the inlet of the hydraulic motor located at station 2. The pressure at station 1 in the hydraulic tank is atmospheric.

Problem 3a

Solve the following problem on the same lines as above (problem to be solved by participants).

In the above hydraulic system, following data is available:

> (a) The pump adds a power of 4 hp to the fluid
> (b) Pump flow rate is 25 gpm

(c) The pipe diameter is 20 mm
(d) The specific gravity of oil is 0.9
(e) The head loss between the two stations is 40 ft of oil.

Find the pressure available at the inlet of the motor at station 2.

Sample Problem 4

Designing of a heat exchanger: the following example will show us how to calculate the rise in temperature of the fluid as it flows through a restriction such as a pressure relief valve, using the relation,

$$\text{Temperature rise} = \frac{\text{Heat generated (Btu/min)}}{\text{(Specific heat of oil} \times \text{oil flow rate)}}$$

Problem:
Oil at 120 °F and 1000 psi is flowing through a pressure relief valve at a rate of 10 gpm. What is the downstream oil temperature?

Specific heat of oil is 0.42 Btu/lb/°F
Oil flow rate of 1 lb/min equals 7.42 gpm.

Problem 4a

Solve the following problem on the same lines as above (problem to be solved by participants).

Oil flows through a pressure relief valve at 15 gpm. Given,

(a) Oil temperature = 130 °F and
(b) Pressure = 2000 psi.

Calculate the rise in temperature of the oil as it flows through the relief valve.

Sample Problem 5

This example deals with the sizing of a heat exchanger in a hydraulic system.

Problems:
A hydraulic pump operates at 100 psi and delivers oil to a hydraulic actuator. Oil discharges through the pressure relief valve (PRV) during 50% of the cycle time. The pump has an overall efficiency of 85% and 10% of the power is lost due to frictional pressure losses in the lines. What is the heat exchanger rating required to dissipate the generated heat?

Problem 5a

Solve the following problem on the same lines as above (problem to be solved by participants).

A hydraulic pump operates at 2000 psi and delivers oil at 15 gpm to a hydraulic actuator. Oil discharges through the PRV, 60% of the cycle time. The pump has an overall efficiency of 82% and 10% of the power is lost due to frictional losses in the lines. What rating heat exchanger is required to dissipate the heat generated?

Appendix B

Practical solutions

Exercise 1

1.1 Calculate the pressure due to a column of 0.5 m of

  (a) Water
  (b) Oil of specific gravity 0.8
  (c) Mercury of specific gravity 13.6.

Solution: Height of liquid column 'h' = 0.5 m
  S_{oil} = 0.8
  $S_{mercury}$ = 13.6
  Density of water (ρ) = 1000 kg/m^3
  Gravitational force (g) = 9.8 m/s^2

  (a) We know according to hydrostatic law that pressure developed in a fluid due to a height 'h' is given by

$$P = \rho g h$$
$$\Rightarrow 1000 \times 9.8 \times 0.5 = 4900 \ \text{N/m}^2$$

  ∴ *Pressure due to a column of 0.5 m of water = 4900 N/m^2 (SI units)*
  [In metric units = 4900/9.8 × 100 × 100 = 0.05 kg/cm^2 (where 9.81 is to convert Newton into kgf and 100 × 100 is to convert m^2 into cm^2).]

  (b) Pressure due to oil column: $P_{(oil)} = \rho_{(oil)} \, gh$
    We know,

$$\text{Specific gravity of oil } (S_{oil}) = \frac{\text{Density of oil}}{\text{Density of water}}$$
$$\Rightarrow 0.8 = \rho_{oil} / 1000 \Rightarrow \rho_{oil} = 0.8 \times 1000 = 800 \ \text{kg/m}^3$$
$$\therefore P_{oil} = 800 \times 9.8 \times 0.5 = 3920 \ \text{N/m}^2$$

Pressure due to oil column = 3920 N/m^2.

(c) Pressure due to mercury column: $P_{(mercury)} = \rho_{(mercury)} \, gh$

$$\rho_{(mercury)} = S_{mercury} \times \rho_{(water)}$$
$$\Rightarrow 13.6 \times 1000 = 13\,600 \text{ kg/m}^3$$
$$\therefore P_{(mercury)} = 13\,600 \times 9.8 \times 0.5 = 66\,640 \text{ N/m}^2.$$

Pressure due to mercury column = 66 640 N/m².

1.2 The intensity of pressure at a point in a fluid is 4 N/cm². Find the corresponding height of fluid when the fluid is

(a) Water
(b) Oil of specific gravity is 0.9.

Solution: $P = 4 \text{ N/cm}^2 = 40\,000 \text{ N/m}^2$
$$S_{oil} = 0.9$$
$$\rho_{water} = 1000 \text{ kg/m}^3$$

(a) $P = \rho g h$
$$\Rightarrow 40\,000 = 1000 \times 9.81 \times h$$
$$\Rightarrow 40\,000 = 9810\,h$$
$$\Rightarrow h = 4.07 \text{ m}.$$

This means that the height of water column required to produce a pressure of 4 N/cm² is 4.07 m.

(b) $S = \rho_{oil} / \rho_{water}$
$$\Rightarrow \rho_{oil} = 0.9 \times 1000 = 900 \text{ kg/m}^3$$
Since,
$$P = \rho g h$$
$$\Rightarrow 40\,000 = 900 \times 9.81 \times h$$
$$\Rightarrow 40\,000 = 8829\,h$$
$$\Rightarrow h = 4.53 \text{ m}.$$

This shows that the height of mercury column required to produce a pressure of 4 N/cm² is 4.53, whereas using a water column of height 4.07 m can produce the same pressure.

1.3 A bullet of mass 8 g travels at a speed of 400 m/s. It penetrates a target, which offers a constant resistance of 1000 N to the motion of the bullet. Calculate

(a) The kinetic energy of the bullet
(b) The distance through which the bullet has penetrated.

Solution: Mass of bullet = 8 g = 8/1000 kg
Velocity of bullet = 400 m/s
Force = 1000 N

(a) We know kinetic energy = $\frac{1}{2}mv^2$
$$\Rightarrow (1 \times 8 \times 400^2) / (2 \times 1000)$$
$$= 640 \text{ J}$$
$\therefore$ Kinetic energy of the bullet = 640 J

(b) We know that work done = Force × Distance

$$640 = 1000 \times s$$
$$s = 640/1000$$
$$= 0.64 \text{ m}$$

∴ The distance the bullet has penetrated the fixed target = 0.64 m.

1.4 A hydraulic press has a ram of 25 cm diameter and a plunger of 4.0 cm diameter. Find the weight lifted by the hydraulic press when the force applied at the plunger is 400 N.

Solution: Diameter of ram = 25 cm

$\Rightarrow$ Area of ram = $\pi d^2/4$

$\Rightarrow 3.14\,(25)^2/4$

Area of ram $(A) = 490.62$ cm^2

Diameter of plunger = 4 cm

$\Rightarrow$ Area of plunger $(a) = 12.56$ cm^2

Force applied on plunger $(F) = 400$ N

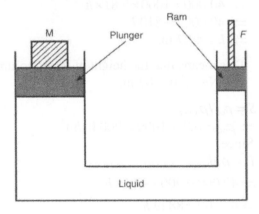

Now the force 'F' applied to the plunger exerts a pressure on the liquid, which is given by

$$P = \frac{F}{A}$$

We know $F = 400$ N, A = Area of the plunger $(a) = 12.56$ cm^2

$$\therefore \ P = \frac{400}{12.56} = 31.9 \text{ N/cm}^2$$

According to Pascal's law this pressure is transmitted to every point in the fluid and also acts on the ram, as a result of which the ram will experience an upward force (W) given by

$$W = P \times A$$

Where 'P' is the pressure exerted by the liquid on the ram and 'A' is the area of the ram.

$$W = 31.9 \times 490.62 = 15\,650.8 \text{ N} = 15.65 \text{ kN}$$

∴ By application of a force of 400 N on the plunger, the ram will be able to lift a weight of 15.65 kN.

1.5 Calculate the gage pressure and absolute pressure at a point 3 m below the free surface of a liquid having a density of 1.53×10^3 kg /m³, if the atmospheric pressure is equal to 750 mm of mercury. The specific gravity of mercury is 13.6 and density of water is 1000 kg/m³.

Solution: $h = 3$ m $\rho_{(liquid)} = 1.53 \times 10^3$ kg/m³

Pressure head (atmospheric) = 750 mm of Hg

$S_{(mercury)} = 13.6$ kg/m³

$\rho_{(water)} = 1000$ kg/m³

We know, Absolute pressure = Gage pressure + Atmospheric pressure.

(a) Gage pressure = $\rho_{(liquid)} gh$

$\Rightarrow 1.53 \times 10^3 \times 9.81 \times 3 = 45\,028$ N/m²

∴ Gage pressure = 45 028 N/m² = 4.5028 N/cm²

Atmospheric pressure = $\rho_{(mercury)} gh$

We know,

$$\rho_{(mercury)} = S_{(mercury)} \times \rho_{(water)}$$
$$\Rightarrow \rho_{(mercury)} = 13.6 \times 1000$$
$$\rho_{(mercury)} = 13\,600 \text{ kg/m}^3$$

We know,

$h = 750$ mm of mercury = 0.750 m.

$\Rightarrow$ Atmospheric pressure = $13\,600 \times 9.81 \times 0.750 = 100\,062$ N/m²

∴ Atmospheric pressure = 10.0062 N/cm²

Hence (a) becomes

$$4.50 + 10.0 = 14.50 \text{ N/cm}^2$$

∴The absolute pressure is 14.50 N/cm² and Gage pressure is 4.50 N/cm².

1.6 The diameters of a pipe are 15 cm and 20 cm at sections 1 and 2 respectively. The velocity of water flowing through the pipe is 4 m/s at section 1. Determine the discharge through the pipe and also the velocity of flow at section 2.

Solution: $d_1 = 15$ cm

$v_1 = 4$ m/s = 400 cm/s (section 1.1)

$d_2 = 20$ cm

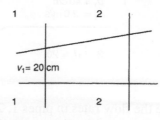

We know from continuity equation

$$A_1 v_1 = A_2 v_2 \qquad\qquad (A.1)$$

$$A_1 = \frac{(\pi d_1^2)}{4}$$
$$= \frac{\{3.14 \times (15^2)\}}{4}$$
$$= 176.63 \text{ cm}^2$$

$$A_2 = \frac{(3.14 \times 20 \times 20)}{4}$$
$$= 314 \text{ cm}^2$$

∴ (A.1) becomes

$$176.63 \times 400 = 314 \times v_2$$

$$\Rightarrow v_2 = \frac{(176.63 \times 400)}{314}$$
$$= 225 \text{ cm/s}$$
$$= 22.5 \text{ m/s}$$

We know

$$\text{Discharge/flow rate } (Q) = A_1 v_1 = A_2 v_2$$
$$\Rightarrow Q = 176.63 \times 400 = 70\,652 \text{ cm}^3/s$$
$$\Rightarrow Q = 0.071 \text{ m}^3/s$$

Thus it is seen that water in flowing from section 1 to 2, loses velocity, which is quite obvious considering the increase in diameter. (However in doing so, it must be understood that there will be a rise in the pressure energy.)

1.7   A pipe of 40 cm diameter carrying water branches into another 2 pipes of diameters 25 and 20 cm respectively. If the average velocity in the 40 cm pipe is 2.0 m/s find the flow rate in this pipe. Also determine the velocity in the 20 cm pipe if the average velocity in the 25 cm pipe is 1.8 m/s.

Solution:

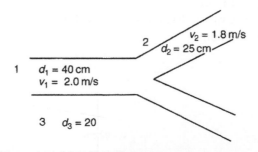

Let Q_1, Q_2 and Q_3 be the flow rates in pipes 1, 2 and 3 respectively.
(Note that the continuity equation is not applicable here, because the fluid in pipe 1 is getting divided or removed.)

We know,

$$Q_1 = Q_2 + Q_3 \qquad (A.2)$$

Discharge (Q_1) through pipe $1 = A_1 v_1$

Where

$$A_1 = 3.14 \times \frac{\left(\dfrac{40}{100}\right)^2}{4}$$

$$= 0.1256 \text{ m}^2$$

$$\Rightarrow Q_1 = 0.1256 \times 2 = 0.251 \text{ m}^3/s$$

$$Q_2 = A_2 v_2$$

$$\Rightarrow \frac{\left(3.14 \times 0.25^2\right)}{4} \times 1.8 = 0.088 \text{ m}^3/s$$

$$\Rightarrow Q_2 = 0.088 \text{ m}^3/s$$

$\therefore$ (A.2) becomes,

$$0.251 = 0.088 + Q_3$$

$$Q_3 = 0.251 - 0.088 = 0.163 \text{ m}^3/s$$

$$\Rightarrow Q_3 = 0.163 \text{ m}^3/s$$

Now,

$$Q_3 = A_3 v_3 \Rightarrow v_3 = \frac{0.163}{\dfrac{\left(3.14 \times 0.20^2\right)}{4}} \Rightarrow v_3 = 5.19 \text{ m/s}$$

Exercise 2

A few problems and their solutions, which deal with, the designing and analyzing simple circuits in a hydraulic system.

Sample Problem 1

Deals with air to hydraulic pressure booster system: The following exercise will tell us how to calculate the load-carrying capacity in a pressure booster system.

Problem:
The figure shows a pressure booster system used to drive a load '*F*' via a hydraulic cylinder. The following data is available.

Inlet air pressure $(P_1) = 100$ psi
Air piston Area $(A_1) = 20$ sq.in.
Oil piston area $(A_2) = 1$ sq.in.
Load piston area $(A_3) = 25$ sq.in.

Find the load-carrying capacity of the system.

Solution:
First, find the booster discharge pressure (P_2):

$$\text{Booster input force} = \text{Booster output force}$$

$$P_1 A_1 = P_2 A_2$$

So

$$P_2 = \frac{(P_1 A_1)}{A_2}$$

$$= \frac{(100 \times 20)}{1}$$

$$= 2000 \text{ psi}$$

As per Pascal's law,

$$P_3 = P_2 = 2000 \text{ psi}$$
$$\text{Force } (F) = P_3 \times A_3$$
$$= 2000 \times 25$$
$$= 20\,000 \text{ lb}$$

Assuming an air pressure of 100 psi without a booster, the area of the piston could have been as high as 500 sq.in. This is how a designer can actually reduce the cylinder size, by using a pressure booster system.

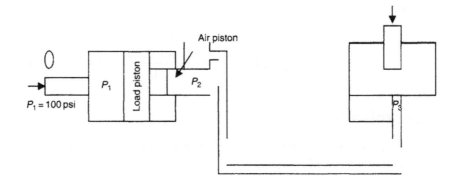

Sample Problem 2

Deals with the hydraulic horsepower analysis technique.

Problem:
A hydraulic cylinder is used to compress a car body down to a bale size in 10 s. The operation requires a 10 ft stroke (S) and a force (F_{load}) of 8000 lb. If a 1000 psi pressure (P) pump is selected, find

    (a) The required piston area
    (b) The necessary pump flow rate
    (c) The hydraulic horsepower delivered by the cylinder.

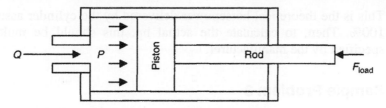

Solution:

Here F_{load} is the force required to crush the car for which the pump used can deliver a pressure of 1000 psi. So, to get the area of the piston required to take this load,

(a) Force $= P \times A$

so

$$A = \frac{F_{load}}{P}$$
$$= \frac{8000}{1000}$$
$$= 8 \text{ sq.in.}$$

(b) The volumetric displacement of the cylinder equals the fluid volume swept by the cylinder during its stroke length (S) while the required pump flow rate equals the volumetric displacement divided by the time required for the stroke. So,

$$Q \text{ (ft}^3/\text{s)} = \frac{(A \times S)}{t}$$
$$= \frac{((8/144) \times 10)}{10}$$
$$= 0.056 \text{ ft}^3/\text{s}$$
$$1 \text{ ft}^3/\text{s} = 448 \text{ gpm}$$

so

$$Q = 448 \times 0.056$$
$$= 25.1 \text{ gpm}$$

In order to calculate the power delivered we will use the equation

$$Hp = \frac{\{P \text{ (psi)} \times Q \text{ (gpm)}\}}{1714}$$

This has been derived by using the conversion factors, keeping in view the basic power-energy equation which is

$$\text{Power} = \frac{\text{Energy}}{\text{Time}}$$

So

$$Hp = \frac{(1000 \times 25.1)}{1714}$$
$$= 14.6 \text{ hp}$$

This is the theoretical horsepower delivered by the cylinder assuming its efficiency to be 100%. Then, to calculate the actual hp, this should be multiplied by the efficiency specified by the manufacturer.

Sample Problem 3

In the following example, we shall study how a hydraulic system problem is solved using the metric-SI system units.

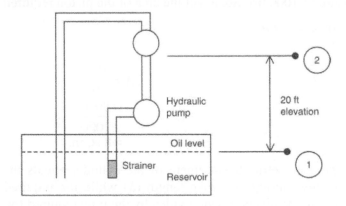

Problem:

For the above hydraulic system,

1. The pump adds a power of 5 hp (3730 W) to the fluid
2. Pump flow rate is 0.001896 m^3/s (113 lph)
3. The pipe diameter is 25.4 mm (0.0254 m)
4. The specific gravity of oil is 0.9
5. The elevation difference between station 1 and 2 is 20 ft (6.096 m)
6. The head loss between the two stations is 9.144 m of oil.

Find the pressure available at the inlet of the hydraulic motor located at station 2. The pressure at station 1 in the hydraulic tank is atmospheric.

Solution:

Applying Bernoulli's equation to the two stations,

$$\frac{Z_1 + P_1}{\gamma} + \frac{V_1^2}{2g} + H_p - H_m - H_L = Z_2 + \frac{P_2}{\gamma} + \frac{V_2^2}{2g}$$

Since there is no hydraulic motor between the two stations, $H_m = 0$. Also $V_1 = 0$ as the cross-section of the reservoir is large.

$$Z_2 - Z_1 = 6.096 \text{ m} \quad \text{and} \quad P_1 = 0$$

Substituting the known values we get,

$$Z_1 + 0 + 0 + H_p - 0 - 9.144 = Z_2 + \frac{P_2}{\gamma} + \frac{V_2^2}{2g}$$

Solving for $\dfrac{P_2}{\gamma}$, we have

$$\frac{P_2}{\gamma} = (Z_1 - Z_2) + H_p - \frac{V_2^2}{g} - 9.144$$

$$= 6.096 + H_p - \frac{V_2^2}{2g} - 9.144$$

$$= H_p - \frac{V_2^2}{2g} - 15.24 \qquad\qquad (A.3)$$

To calculate the H_p (head developed) by the pump, we have equation:

$$H_p \text{ (ft)} = \frac{\{3950 \times \text{Power (hp)}\}}{Q \text{ (gpm)} \times S_g}$$

We know that:

$$H_p \text{ (ft)} = 3.28 \times H_p \text{ (m)}$$

$$Q \text{ (gpm)} = \frac{1}{0.0000632 \text{ m}^3/\text{s}}$$

After converting the units required in the above equation for the known parameters we have equation:

$$H_p \text{ (m)} = \frac{\{0.250 \times \text{hp}\}}{3.28 \times Q \text{ (m}^3/\text{s)} \times S_g}$$

$$= \frac{(0.250 \times 5)}{(3.28 \times 0.001896 \times 0.9)}$$

$$= 223.3 \text{ m}$$

To find the velocity V_2, we have the equation

$$Q = A \times V$$

So

$$V_2 = \frac{Q (\text{m}^3/\text{s})}{A (\text{m}^2)}$$

$$= \frac{0.001896}{\left\{\left(\dfrac{\Pi}{4}\right) \times (0.0254)^2\right\}}$$

$$= 3.74 \text{ m/s}$$

$$\frac{V_2^2}{2g} = \frac{(3.74)^2}{2 \times 9.81}$$

$$= 0.714 \text{ m}$$

Now equating the above values in (A.3)

$$\frac{P_2}{\gamma} = 223.3 - 0.714 - 15.24$$

$$= 207.3 \text{ m}$$

$$\gamma = S_g \times \gamma_{water}$$

$$= 0.9 \times 9797 \text{ (N/m}^3)$$

$$= 8817 \text{ N/m}^3$$

So

$$P_2 = 207.3 \text{ (m)} \times 8817 \text{ (N/m}^3)$$

$$= 1827764.1 \text{ N/m}^2 \text{ (Pascal)}$$

$$1 \text{ Pascal} = 1000 \text{ kPa}$$

$$P_2 = 1827.764 \text{ kPa}$$

$$1 \text{ Pa} = 0.000145 \text{ psi}$$

So

$$P_2 = 265 \text{ psi}$$

The pressure available at (P_2), at the inlet of the hydraulic motor at station 2 is 265 psi.

Sample Problem 4

Designing of a heat exchanger: The following example will show us how to calculate the rise in temperature of the fluid as it flows through the restriction such as a pressure relief valve. The relation is used to calculate this.

$$\text{Temperature rise} = \frac{\text{Heat generated (Btu/min)}}{\text{(Specific heat of oil} \times \text{oil flow rate)}}$$

Problem:
Oil at 120 °F and 1000 psi is flowing through a pressure relief valve at 10 gpm. What is the downstream oil temperature?

Specific heat of oil is 0.42 Btu/lb/ °F
Oil flow rate of 1 lb/min equals 7.42 gpm.

Solution:
Initially, the horsepower lost should be calculated and then converted in terms of heat generated in Btu/min.

$$\text{Hp} = \frac{\{P \text{ (psi)} \times Q \text{ (gpm)}\}}{1714}$$

$$= \frac{1000 \times 10}{1714}$$

$$= 5.83$$

$$= 247 \text{ Btu/min}$$

As we have 1 hp = 42.4 Btu/min, so to calculate the rise in temperature in the system we need to have all the parameters measured in the same units.

$$\text{Oil flow rate in lb/min} = 7.42 \times 10$$
$$= 74.2 \text{ lb/min}$$

Substituting the values in the equation:

$$Q \text{ (Btu/min)} = m \text{ (flow rate in lb/min)} \times Cp \text{ (specific heat)} \times \text{Rise in temperature}$$
$$247 = 74.2 \times 0.42 \times \text{temperature rise}$$

So temperature rise = 7.9°F

When sizing a heat exchanger, the heat load value for the entire system is calculated in the units of Btu/h. As a thumb rule, the relation that can be used for this is:
1 hp equals to 2544 Btu/h of heat generation.

Sample Problem 5

This example deals with the sizing of a heat exchanger in a hydraulic system.

Problems:
A hydraulic pump operates at 100 psi and delivers oil to a hydraulic actuator. Oil discharges through the pressure relief valve (PRV) during 50% of the cycle time. The pump has an overall efficiency of 85%, and 10% of the power is lost due to frictional pressure losses in the lines. In order to dissipate all the generated heat, calculate the heat exchanger rating.

Solution:
We have the equation Hp = {P (psi) × Q (gpm)}/1714. Using this equation calculate the hp loss at various stages in the system.

1. Hp loss at the pump: The overall efficiency of the pump is 85%, so the hp loss in the pump amount to 15%.

$$\text{Hp}_{\text{loss}} \text{ in pump} = \left\{ \frac{(1000 \times 20)}{1714} \right\} \times 0.15$$
$$= 1.75$$

2. Hp loss in the PRV is to be taken only 50% of the total cycle

$$\text{Hp}_{\text{loss}} \text{ in the PRV} = \left\{ \frac{(1000 \times 20)}{1714} \right\} \times 0.50$$
$$= 5.83$$

3. The losses in the lines amount to 10% of losses generated by the flow in the PRV

$$= 5.83 \times 0.10$$
$$= 0.583$$

$$\text{Total Hp loss} = 1.75 + 5.83 + 0.58$$
$$= 8.16$$

We know that 1 hp = 2544 Btu/h

So

$$\text{The rating of the heat exchanger} = 8.16 \times 2544$$
$$= 20\ 759\ \text{Btu/h}$$

The heat exchanger should be designed to dissipate 20 759 Btu/h of heat from the system.

Index

THIS BOOK WAS DEVELOPED BY IDC TECHNOLOGIES

WHO ARE WE?

IDC Technologies is internationally acknowledged as the premier provider of practical, technical training for engineers and technicians.

We specialise in the fields of electrical systems, industrial data communications, telecommunications, automation & control, mechanical engineering, chemical and civil engineering, and are continually adding to our portfolio of over 60 different workshops. Our instructors are highly respected in their fields of expertise and in the last ten years have trained over 50,000 engineers, scientists and technicians.

With offices conveniently located worldwide, IDC Technologies has an enthusiastic team of professional engineers, technicians and support staff who are committed to providing the highest quality of training and consultancy.

TECHNICAL WORKSHOPS

TRAINING THAT WORKS

We deliver engineering and technology training that will maximise your business goals. In today's competitive environment, you require training that will help you and your organisation to achieve its goals and produce a large return on investment. With our "Training that Works" objective you and your organisation will:

- Get job-related skills that you need to achieve your business goals
- Improve the operation and design of your equipment and plant
- Improve your troubleshooting abilities
- Sharpen your competitive edge
- Boost morale and retain valuable staff
- Save time and money

EXPERT INSTRUCTORS

We search the world for good quality instructors who have three outstanding attributes:

1. Expert knowledge and experience – of the course topic
2. Superb training abilities – to ensure the know-how is transferred effectively and quickly to you in a practical hands-on way
3. Listening skills – they listen carefully to the needs of the participants and want to ensure that you benefit from the experience

Each and every instructor is evaluated by the delegates and we assess the presentation after each class to ensure that the instructor stays on track in presenting outstanding courses.

HANDS-ON APPROACH TO TRAINING

All IDC Technologies workshops include practical, hands-on sessions where the delegates are given the opportunity to apply in practice the theory they have learnt.

REFERENCE MATERIALS

A fully illustrated workshop book with hundreds of pages of tables, charts, figures and handy hints, plus considerable reference material is provided FREE of charge to each delegate.

ACCREDITATION AND CONTINUING EDUCATION

Satisfactory completion of all IDC workshops satisfies the requirements of the International Association for Continuing Education and Training for the award of 1.4 Continuing Education Units.

IDC workshops also satisfy criteria for Continuing Professional Development according to the requirements of the Institution of Electrical Engineers and Institution of Measurement and Control in the UK, Institution of Engineers in Australia, Institution of Engineers New Zealand, and others.

CERTIFICATE OF ATTENDANCE

Each delegate receives a Certificate of Attendance documenting their experience.

100% MONEY BACK GUARANTEE

IDC Technologies' engineers have put considerable time and experience into ensuring that you gain maximum value from each workshop. If by lunch time of the first day you decide that the workshop is not appropriate for your requirements, please let us know so that we can arrange a 100% refund of your fee.

ONSITE WORKSHOPS

All IDC Technologies Training Workshops are available on an on-site basis, presented at the venue of your choice, saving delegates travel time and expenses, thus providing your company with even greater savings.

OFFICE LOCATIONS

AUSTRALIA · CANADA · IRELAND · NEW ZEALAND · SINGAPORE · SOUTH AFRICA · UNITED KINGDOM · UNITED STATES

idc@idc-online.com • www.idc-online.com

Printed and bound by CPI Group (UK) Ltd, Croydon, CR0 4YY

03/10/2024

01040338-0015